Collins

CONCISE REVISION COURSE

CSEC® Chemistry

Anne Tindale

Collins

HarperCollins Publishers Ltd
The News Building
1 London Bridge Street
London SE1 9GF

HarperCollins *Publishers*
Macken House, 39/40 Mayor Street Upper,
Dublin 1, D01 C9W8, Ireland

First edition 2016

10 9

ISBN 978-0-00-815788-3

www.collins.co.uk/caribbeanschools

A catalogue record for this book is available from the British Library.

Typeset by QBS
Printed in India

Author: Anne Tindale
Publisher: Elaine Higgleton
Commissioning Editor: Peter Dennis
Managing Editor: Sarah Thomas
Copy Editor: Delphine Phin
Editor: Aidan Gill
Proofreader: Tim Jackson
Artwork: QBS
Cover: Kevin Robbins and Gordon MacGilp

Acknowledgements

The publishers would like to thank the following for permission to use their photos in this book:
p72: Anne Tindale; p89: Anne Tindale; p103: Andre Nitsievsky/Shutterstock; p134: Photo smile/Shutterstock; p146l: Toa55/Shutterstock; p146r: Lodimup/Shutterstock; p167l: PChStudios/Shutterstock; p167r: Mark Sykes/Science Photo Library; p178: Mikadun/Shutterstock; p184l: ThamKC/Shutterstock; p184r: Mexrix/Shutterstock; p186: Anne Tindale; p187: Anne Tindale

Contents

The pathway to success

About this book

This book has been written primarily as a **revision course** for students studying for the CSEC® Chemistry examination. The facts are presented **concisely** using a variety of formats which makes them **easy to understand** and **learn.** Key words are highlighted in **bold** type and important definitions which must be learnt are written in *italics* and highlighted in colour. **Annotated diagrams** and **tables** have been used wherever possible and **worked examples** have been included where appropriate. **Questions** to help test knowledge and understanding, and provide practice for the actual examination, are included throughout the book.

The following sections provide **valuable information** on the format of the CSEC® examination, how to revise successfully, successful examination technique, key terms used on examination papers and School-Based Assessment.

The CSEC® Chemistry syllabus and this book

The **CSEC® Chemistry syllabus** is available online at **http://cxc-store.com**. You are strongly advised to read through this syllabus carefully since it provides detailed information on the specific objectives of each topic of the course, School-Based Assessment (SBA) and the format of the CSEC® Examination. Each chapter in **this book** covers a particular topic in the syllabus.

- **Chapters 1 to 13** cover topics in Section A, **Principles of Chemistry**
- **Chapters 14 to 17** cover topics in Section B, **Organic Chemistry**
- **Chapters 18 to 24** cover topics in Section C, **Inorganic Chemistry**

At the end of each chapter, or section within a chapter, you will find a selection of **revision questions**. These questions test your **knowledge** and **understanding** of the topic covered in the chapter or section. At the end Chapters 13, 17 and 24 you will find a selection of **exam-style questions** which also test how you **apply** the knowledge you have gained and help prepare you to answer the different styles of questions that you will encounter in your CSEC® examination. You will find the answers to all these questions online at **www.collins.co.uk/caribbeanschools**.

The format of the CSEC® Chemistry examination

The examination consists of **two papers** and your performance is evaluated using the following three profiles:

- **Knowledge and comprehension**
- **Use of knowledge**
- **Experimental skills**

Paper 01 (1 ¼ hours)

Paper 01 consists of **60 multiple choice questions**. Each question is worth **1 mark**. Four **choices** of answer are provided for each question of which one is correct.

- Make sure you read each question **thoroughly**; some questions may ask which answer is **incorrect**.
- If you don't know the answer, try to work it out by **eliminating** the incorrect answers. Never leave a question unanswered.

Paper 02 (2 ½ hours)

Paper 02 is divided into **Sections A** and **B**, and consists of **six compulsory questions**, each divided into several parts. Take time to **read the entire paper** before beginning to answer any of the questions.

- **Section A** consists of **three** compulsory **structured questions** whose parts require short answers, usually a word, a sentence or a short paragraph. The answers are to be written in **spaces** provided on the paper. These spaces indicate the length of answer required and answers should be restricted to them.
  - Question 1 is a **data-analysis question** which is worth **25 marks**. The first part usually asks you to take readings from a measuring instrument, such as a set of thermometers, and record these readings in a table. You may then be asked to draw a graph using the information in the table and may be asked questions about the graph or be asked to perform certain calculations. The second part will possibly test your knowledge of tests to identify cations, anions and gases, and there may be a third part which tests your planning and designing skills.
  - Questions 2 and 3 are each worth **15 marks**. They usually begin with some kind of stimulus material, such as a diagram or a table, which you will be asked questions about.
- **Section B** consists of **three** compulsory **extended-response questions**, each worth **15 marks**. These questions require a greater element of **essay** writing in their answers than those in Section A.

The marks allocated for the different parts of each question are clearly given. A total of **100 marks** is available for Paper 02 and the time allowed is **150 minutes**. You should allow about 35 minutes for the data-analysis question worth 25 marks and allow about 20 minutes for each of the other questions. This will allow you time to read the paper fully before you begin and time to check over your answers when you have finished.

Successful revision

The following should provide a guide for **successful revision**.

- **Begin your revision early**. You should start your revision at least two months before the examination and should plan a **revision timetable** to cover this period. Plan to revise in the evenings when you don't have much homework, at weekends, during the Easter vacation and during study leave.

- When you have a **full day** available for revision, consider the day as three sessions of about three to four hours each, **morning, afternoon** and **evening.** Study during two of these sessions only, do something non-academic and relaxing during the third.
- **Read through the topic** you plan to learn to make sure you **understand** it before starting to learn it; understanding is a lot safer than thoughtless learning.
- Try to understand and learn **one topic** in each revision session, more if topics are short and fewer if topics are long.
- **Revise every topic** in the syllabus. Do not pick and choose topics since **all questions** on your exam paper are **compulsory**.
- **Learn the topics in order**. When you have learnt **all** topics **once**, go back to the first topic and begin again. Try to cover each topic **several times.**
- **Revise in a quiet location** without any form of distraction.
- **Sit up to revise**, preferably at a table. Do not sit in a comfy chair or lie on a bed where you can easily fall asleep.
- Obtain copies of **past CSEC® Chemistry examination papers** and use them to practise answering exam-style questions, starting with the most recent papers. These can be purchased online from the CXC Store.
- You can use a variety of different **methods** to **learn** your work. Chose which ones work best for you.
  - **Read the topic several times**, then close the book and try to write down the **main points**. Do not try to memorise your work word for word since work learnt by heart is not usually understood and most questions test **understanding**, not just the ability to repeat facts.
  - **Summarise** the **main points** of each topic on **flash cards** and use these to help you study.
  - **Draw simple diagrams** with **annotations, flow charts** and **spider diagrams** to summarise topics in visual ways which are easy to learn.
  - **Practise drawing** and **labelling** simple line diagrams of **apparatus** you have encountered. You may be asked to reproduce these, e.g. the apparatus used for fractional distillation.
  - **Practise writing equations**. Do not try to learn equations by heart; instead, understand and learn **how** to write and balance them. For example, if you learn that a carbonate reacts with an acid to form a salt, carbon dioxide and water, you can write the equation for **any** carbonate reacting with **any** acid.
  - **Use memory aids** such as:
    - **acronyms**, e.g. **OIL RIG** for oxidation and reduction in terms of electrons; **o**xidation **i**s **l**oss, **r**eduction **i**s **g**ain.
    - **mnemonics**, e.g. '**P**eter **s**ometimes **c**ollects **m**oney **a**t **z**oos **i**n **L**ondon **h**elping **c**razy **m**onkeys and **s**illy **g**iraffes' for the order of metals in the reactivity series; **p**otassium, **s**odium, **c**alcium, **m**agnesium, **a**luminium, **z**inc, **i**ron, (**h**ydrogen), **c**opper, **m**ercury, **s**ilver, **g**old.
    - **associations between words**, e.g. **an**ions - **n**egative (therefore cations must be positive).
  - **Test yourself** using the questions throughout this book and others from past CSEC® examination papers.

Successful examination technique

- **Read the instructions** at the start of each paper very carefully and do **precisely** what they require.
- **Read through the entire paper** before you begin to answer any of the questions.
- **Read each question at least twice** before beginning your answer to ensure you **understand** what it asks.
- **Underline the important words** in each question to help you answer precisely what the question is asking.
- **Re-read** the question when you are **part way through** your answer to check that you are answering what it asks.
- **Give precise** and **factual answers.** You will not get marks for information which is 'padded out' or irrelevant. The number of marks awarded for each answer indicates how long and detailed it should be.
- **Use correct scientific terminology** throughout your answers.
- **Balance all chemical equations** and ensure that you give the correct **state symbols,** especially in ionic equations.
- **Show all working** and give clear statements when answering questions that require **calculations.**
- Give every **numerical answer** the appropriate **unit** using the proper abbreviation/symbol e.g. cm^3, g, °C.
- If a question asks you to give a **specific number of points**, use **bullets** to make each separate point clear.
- If you are asked to give **similarities** and **differences**, you must make it clear which points you are proposing as similarities and which points as differences. The same applies if you are asked to give **advantages** and **disadvantages.**
- **Watch the time** as you work. Know the time available for each question and stick to it.
- **Check over your answers** when you have completed all the questions, especially those requiring calculations.
- **Remain in the examination room** until the **end** of the examination and recheck your answers again if you have time to ensure you have done your very best. Never leave the examination room early.

Some key terms used on examination papers

Account for: provide reasons for the information given.

Calculate: give a numerical solution which includes all relevant working.

Compare: give similarities and differences.

Construct: draw a graph or table using data provided or obtained.

Contrast: give differences.

Deduce: use data provided or obtained to arrive at a conclusion.

Define: state concisely the meaning of a word or term.

Describe: provide a detailed account which includes all relevant information.

Determine: find a solution using the information provided, usually by performing a calculation.

Discuss: provide a balanced argument which considers points both for and against.

Distinguish between or **among**: give differences.

Evaluate: determine the significance or worth of the point in question.

Explain: give a clear, detailed account which makes given information easy to understand and provides reasons for the information.

Illustrate: make the answer clearer by including examples or diagrams.

Justify: provide adequate grounds for your reasoning.

Outline: write an account which includes the main points only.

Predict: use information provided to arrive at a likely conclusion or suggest a possible outcome.

Relate: show connections between different sets of information or data.

State or **list**: give brief, precise facts without detail.

Suggest: put forward an idea.

Tabulate: construct a table to show information or data which has been given or obtained.

Drawing tables and graphs

Tables

Tables can be used to record numerical data, observations and inferences. When drawing a table:

- **Neatly enclose** the table and draw vertical and horizontal **lines** to separate columns and rows.
- When drawing **numerical tables**, give the correct column headings which state the **physical quantities** measured and give the correct **units** using proper abbreviations/symbols, e.g. cm^3, g, °C.
- Give the appropriate number of **decimal places** when recording numerical data.
- When drawing **non-numerical tables**, give the correct column headings and **all observations**.
- Give the table an appropriate **title** which must include reference to the responding variable and the manipulated variable.

Graphs

Graphs are used to display numerical data. When drawing a graph:

- Plot the **manipulated variable** on the **x-axis** and the **responding variable** on the **y-axis**.
- Choose appropriate **scales** which are easy to work with and use as much of the graph paper as possible.
- Enter **numbers** along the axes and **label** each axis, including relevant units, e.g. cm^3, g, °C.
- Use a **small dot** surrounded by a small circle to plot each point.
- Plot each point **accurately**.
- Draw a smooth curve or straight line of **best fit** which need not necessarily pass through all the points.
- Give the graph an appropriate **title** which must include reference to the responding variable and the manipulated variable.

School-Based Assessment (SBA)

School-Based Assessment (SBA) is an integral part of your CSEC® examination. It assesses you in the **Experimental Skills** and **Analysis and Interpretation** involved in laboratory and field work, and is worth **20%** of your final examination mark.

- The assessments are carried out in your school by **your teacher** during Terms 1 to 5 of your two-year programme.
- The assessments are carried out during **normal practical classes** and not under examination conditions. You have every opportunity to gain a high score in each assessment if you make a **consistent effort** throughout your two-year programme.
- Assessments will be made of the following **four skills**:
  - Manipulation and Measurement
  - Observation, Recording and Reporting
  - Planning and Designing
  - Analysis and Interpretation

As part of your SBA, you will also carry out an **Investigative Project** during the second year of your two-year programme. This project assesses your **Planning and Designing,** and **Analysis and Interpretation** skills. If you are studying two or three of the single science subjects, Biology, Chemistry and Physics, you may elect to carry out ONE investigation only from any one of these subjects.

You will be required to keep a practical workbook in which you record all of your practical work and this may then be moderated externally by CXC.

1 The states of matter

Chemistry is the study of the composition, structure, properties and reactions of **matter**. Everything around us is made of matter.

Matter *is anything that has volume and mass.*

All matter is made of **particles** and can exist in three different **states**:

- The **solid** state
- The **liquid** state
- The **gaseous** state

The particulate theory of matter

There are four main ideas behind the **particulate theory of matter**:

- All matter is composed of **particles**.
- The particles are in **constant motion** and temperature affects their speed of motion.
- The particles have **empty spaces** between them.
- The particles have **forces of attraction** between them.

Evidence to support the particulate theory of matter

The processes of **diffusion** and **osmosis** provide **evidence** to support the fact that all matter is made of **particles**.

Diffusion

Diffusion *is the net movement of particles from a region of higher concentration to a region of lower concentration, until the particles are evenly distributed.*

Particles in **gases** and **liquids** are capable of diffusing.

Example 1

When pieces of cotton wool soaked in concentrated ammonia solution and concentrated hydrochloric acid are placed simultaneously at opposite ends of a glass tube, a white ring of ammonium chloride forms inside the tube. Ammonia solution gives off ammonia gas and hydrochloric acid gives off hydrogen chloride gas. The particles of the gases **diffuse** through the air inside the tube, collide and react to form ammonium chloride:

$$\text{ammonia} + \text{hydrogen chloride} \longrightarrow \text{ammonium chloride}$$

$$NH_3(g) + HCl(g) \longrightarrow NH_4Cl(s)$$

Ammonia particles diffuse **faster** than hydrogen chloride particles, so the particles collide and react closer to the source of the hydrogen chloride particles.

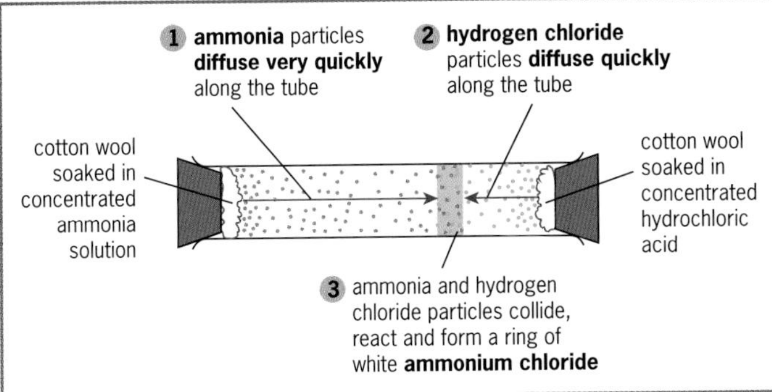

Figure 1.1 *Gases diffuse*

Example 2

When a purple potassium manganate(VII) crystal is placed in water, it **dissolves** to produce a uniformly purple solution. The particles making up the crystal separate from each other and **diffuse** through the spaces between the water particles until they are evenly distributed.

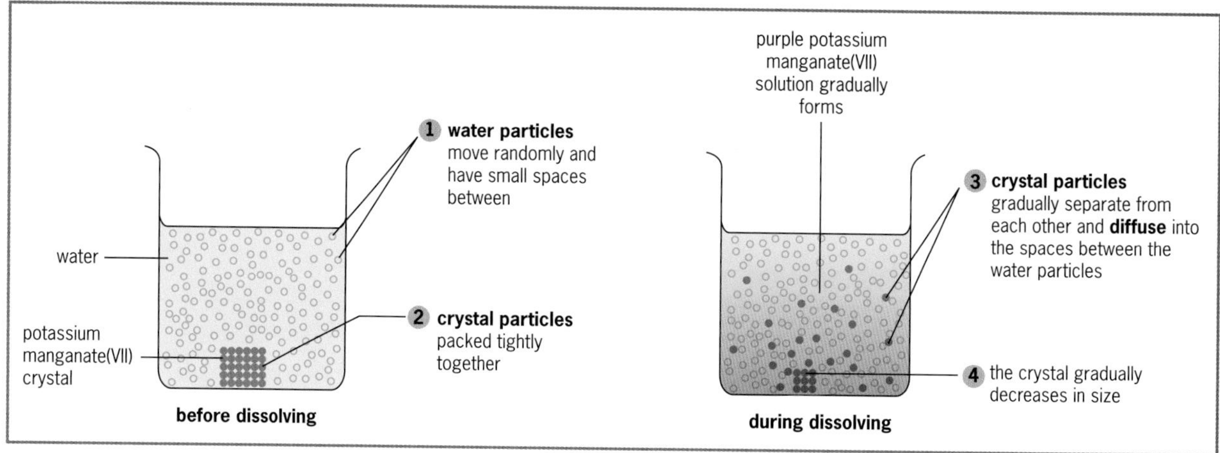

Figure 1.2 *Crystals dissolve*

Osmosis

Osmosis *is the movement of* ***water molecules*** *through a differentially permeable membrane from a solution containing a lot of water molecules, e.g. a dilute solution (or water), to a solution containing fewer water molecules, e.g. a concentrated solution.*

Example 1

When a dilute sucrose solution is separated from a concentrated sucrose solution by a **differentially permeable membrane**, water molecules move through the membrane from the dilute solution into the concentrated solution, but the sucrose molecules cannot move in the other direction. The **volume** of the concentrated solution **increases** and the **volume** of the dilute solution **decreases**.

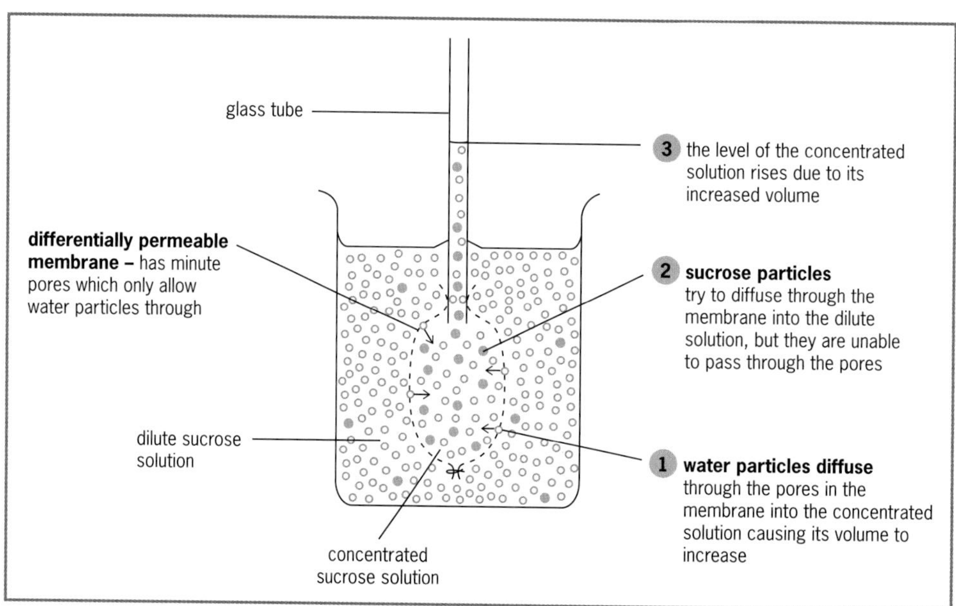

Figure 1.3 *Osmosis explained*

Example 2

The membranes of **living cells** are differentially permeable and the cytoplasm inside the cells contains about 80% water.

- When a strip of living tissue (such as paw-paw) is placed in **water**, water molecules move into the cells by osmosis. Each cell swells slightly, and the strip increases in length and becomes rigid.
- When the strip is placed in a **concentrated sodium chloride solution**, water molecules move out of the cells by osmosis. Each cell shrinks slightly, and the strip decreases in length and becomes softer.

Uses of osmosis

To control garden pests

Slugs and **snails** are garden pests, whose skin is differentially permeable and always moist. When **salt** (sodium chloride) is sprinkled on slugs and snails, it dissolves in the moisture around their bodies forming a concentrated solution. Water inside their bodies then moves out by **osmosis** and into the solution. The slugs and snails die from dehydration if their bodies lose more water than they can tolerate.

To preserve food

Salt and **sugar** are used to preserve foods such as meat, fish and fruit. They both work in the same way:

- They draw water out of the **cells** of the food by osmosis. This prevents the food from decaying because there is no water available in the cells for the chemical reactions which cause the decay.
- They draw water out of **microorganisms** (bacteria and fungi) by osmosis. This prevents the food from decaying because it inhibits the growth of the microorganisms that cause the decay.

Types of particles that make up matter

There are **three** different types of particles that make up matter:

- **Atoms**

  Atoms are the smallest units of a **chemical element** which have all the characteristics of the element. For example, iron is made of iron atoms, Fe (see p. 7).

- **Molecules**

  Molecules are groups of two or more atoms **bonded** together and which can exist on their own. Molecules may be made up of atoms of the same kind, e.g. hydrogen molecules, H_2, are made up of hydrogen atoms, H. Molecules may also be made up of atoms of different kinds, e.g. carbon dioxide molecules, CO_2, are made up of carbon atoms, C, and oxygen atoms, O (see p. 36).

- **Ions**

  Ions are **electrically charged** particles. Ions may be formed from a single atom, e.g. the potassium ion, K^+. They may also be formed from groups of two or more atoms bonded together, e.g. the nitrate ion, NO_3^- (see p. 34).

The three states of matter

The **particulate theory of matter** helps explain the physical properties of matter and the differences between the three states.

Table 1.1 *Comparing the three states of matter*

Property	Solid	Liquid	Gas
Shape	Fixed.	Takes the shape of the part of the container it is in. The surface is always horizontal.	Takes the shape of the entire container it is in.
Volume	Fixed.	Fixed.	Variable – it expands to fill the container it is in.
Density	Usually high.	Usually lower than solids.	Low.
Compressibility	Difficult to compress.	Can be compressed very slightly by applying pressure.	Very easy to compress.
Arrangement of particles	Packed closely together, usually in a regular way:	Have small spaces between and are randomly arranged:	Have large spaces between and are randomly arranged:
Forces of attraction between the particles	Strong.	Weaker than those between the particles in a solid.	Very weak.
Energy possessed by the particles	Possess very small amounts of kinetic energy.	Possess more kinetic energy that the particles in a solid.	Possess large amounts of kinetic energy.
Movement of the particles	Vibrate in their fixed position.	Move slowly past each other.	Move around freely and rapidly.

Changing state

Matter can exist in any of the three states depending on its **temperature**. It can change from one state to another by heating or cooling, as this causes a change in the **kinetic energy** and arrangement of the particles:

- When a solid is **heated**, it usually changes state to a liquid and then to a gas. This occurs because the particles **gain** kinetic energy, move increasingly faster and further apart, and the forces of attraction between them become increasingly weaker.
- When a gas is **cooled**, it usually changes state to a liquid and then to a solid. This occurs because the particles **lose** kinetic energy, move more and more slowly and closer together, and the forces of attraction between them become increasingly stronger.

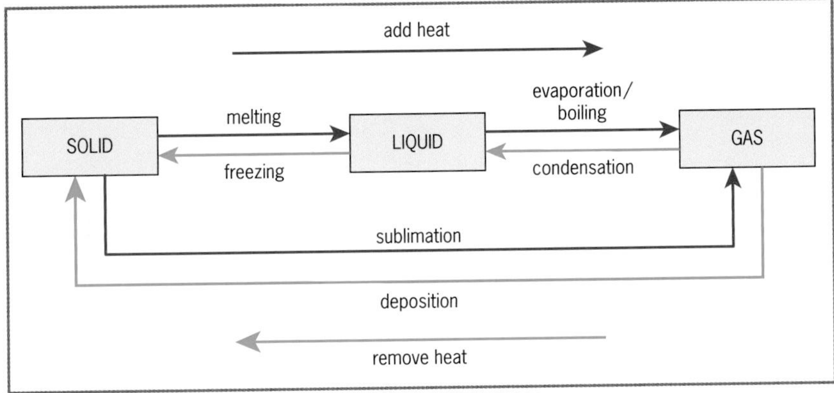

Figure 1.4 *Changing state*

Evaporation and **boiling** are different in the following ways:

- Evaporation can take place at any temperature, whereas boiling occurs at a specific temperature.
- Evaporation takes place at the surface of the liquid only, whereas boiling takes place throughout the liquid.

Substances which **sublimate** (or **sublime**) change directly from a solid to a gas. The reverse process in which a gas changes directly to a solid is called **deposition**. Examples of substances that sublimate include carbon dioxide ('dry ice'), iodine and naphthalene (moth balls).

Heating and cooling curves

- A **heating curve** is drawn when the temperature of a solid is measured at intervals as it is **heated** and changes state to a liquid and then to a gas, and the temperature is then plotted against time.
- A **cooling curve** is drawn when the temperature of a gas is measured at intervals as it is **cooled** and changes state to a liquid and then to a solid, and the temperature is then plotted against time.

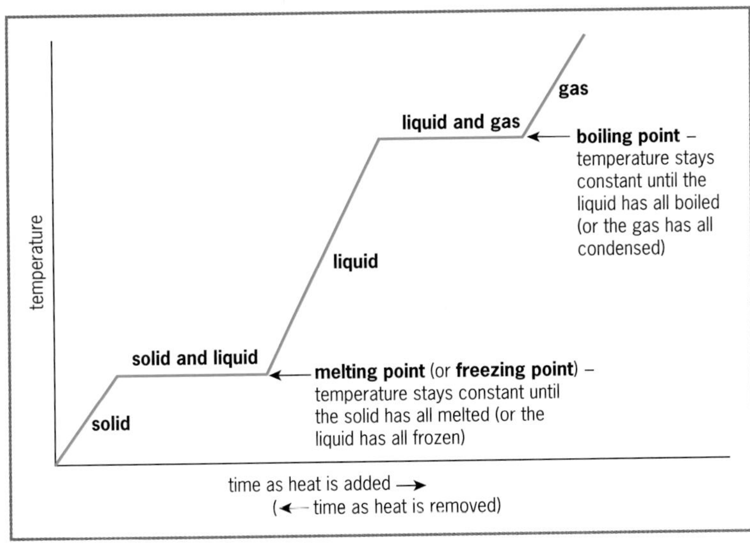

Figure 1.5 *A heating (or cooling) curve*

*The **melting point** is the constant temperature at which a solid changes state into a liquid.*

*The **boiling point** is the constant temperature at which a liquid changes state into a gas.*

*The **freezing point** is the constant temperature at which a liquid changes state into a solid.*

Note The melting and freezing points of any **pure** substance have the same value.

Revision questions

1. Give the FOUR main ideas behind the particulate theory of matter.
2. Define the following:
   a diffusion
   b osmosis
   c melting point
   d boiling point
3. By referring to particles, explain why:
   a when a crystal of red food colouring fell into a beaker of water, after a while all the water became red.
   b when a strip of potato was placed into a concentrated sucrose solution it decreased in length.
4. Explain how salt (sodium chloride) works to preserve fish.
5. By considering the arrangement of particles in each substance, explain why:
   a nitrogen gas is very easy to compress.
   b a solid lump of lead has a fixed shape.
6. Water can exist as solid ice, liquid water and gaseous steam. Explain the difference between these three states in terms of the arrangement of their particles, the movement of their particles, and the forces of attraction between their particles.
7. State TWO differences between evaporation and boiling.
8. What happens when a substance sublimes?

2 Pure substances, mixtures and separations

Matter can be classified into **pure substances** and **mixtures**. These can be further classified as shown:

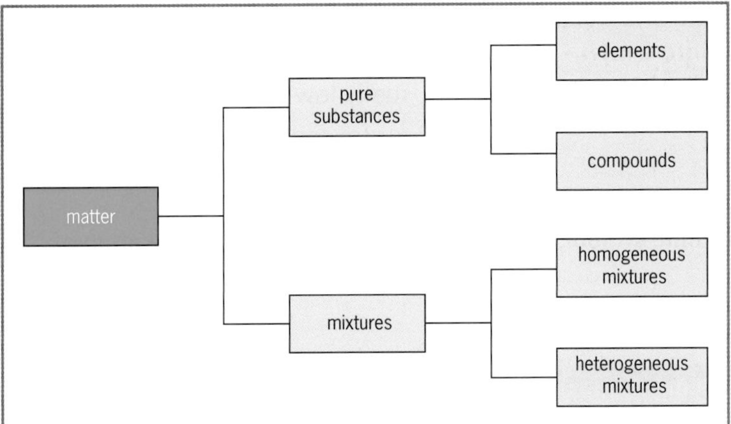

Figure 2.1 *Classification of matter*

Pure substances

*A **pure substance** is composed of a single type of material only.*

Any pure substance possesses certain general **characteristics**:

- Its **composition** is fixed and constant.
- Its **properties** are fixed and constant, for example, its melting point, boiling point and density.
- The **component parts** cannot be separated by any physical process.

To find out if a substance is **pure**, its melting point or boiling point can be measured. If any impurities are present they will usually **lower** its melting point and **raise** its boiling point.

Elements

Elements are the simplest form of matter.

*An **element** is a pure substance that cannot be broken down into simpler substances by using any ordinary physical or chemical process.*

An **atom** is the smallest particle in any element. Each element is composed of atoms of **one kind** only. Most elements are made up of **individual atoms**, e.g. silver (Ag) is made up of individual silver atoms. A few elements are made up of **molecules**, e.g. nitrogen (N_2) is made up of nitrogen molecules, each molecule being composed of two nitrogen atoms.

There are 118 known elements and they can be classified as **metals** or **non-metals**.

Table 2.1 *The physical properties of metals and non-metals compared*

Physical property	Metals	Non-metals
Melting and boiling points	Usually high	Usually low
State at room temperature	Solid (except mercury)	Can be solid, liquid or gas
Appearance of the solid	Shiny	Dull
Bendability of the solid	Malleable (can be hammered into different shapes) and ductile (can be drawn out into wires)	Brittle
Density	Usually high	Usually low
Electrical and thermal conductivity	Good	Poor (except graphite, a form of carbon)

Each element can be represented by an **atomic symbol**. The atomic symbol represents **one atom** of the element.

Table 2.2 *Common metals and their atomic symbols*

Element	Atomic symbol	Element	Atomic symbol
aluminium	Al	lithium	Li
barium	Ba	magnesium	Mg
beryllium	Be	manganese	Mn
calcium	Ca	mercury	Hg
chromium	Cr	nickel	Ni
cobalt	Co	potassium	K
copper	Cu	silver	Ag
gold	Au	sodium	Na
iron	Fe	tin	Sn
lead	Pb	zinc	Zn

Table 2.3 *Common non-metals and their atomic symbols*

Element	Atomic symbol	Element	Atomic symbol
argon	Ar	iodine	I
boron	B	krypton	Kr
bromine	Br	neon	Ne
carbon	C	nitrogen	N
chlorine	Cl	oxygen	O
fluorine	F	phosphorus	P
helium	He	silicon	Si
hydrogen	H	sulfur	S

Note When an atomic symbol consists of two letters, the first letter is always a capital.

Compounds

*A **compound** is a pure substance that is formed from two or more different types of elements which are chemically bonded together in fixed proportions and in a way that their properties have changed.*

Example

sodium + chlorine ⟶ sodium chloride
(element) (element) (compound)

The proportions, by mass, of sodium and chlorine in any pure sample of sodium chloride are always the same and the elements cannot be separated by physical means because they are **chemically bonded** together. The properties of sodium chloride are different from those of both sodium and chlorine.

Compounds can be represented by **chemical formulae**, e.g. the chemical formula for sodium chloride is NaCl and for water it is H_2O.

Mixtures

*A **mixture** consists of two or more substances (elements and/or compounds) which are physically combined together in variable proportions. Each component retains its own individual properties and is not chemically bonded to any other component of the mixture.*

Any mixture possesses certain general **characteristics**:

- Its **composition** can vary.
- Its **properties** are variable because its component parts keep their individual properties.
- Its **component parts** can be separated by physical means (see p. 12).

Homogeneous mixtures

A **homogeneous mixture** is a **uniform** mixture; it has the same composition and properties throughout the mixture. It is not possible to distinguish the component parts from each other. All **solutions** are homogeneous mixtures.

Heterogeneous mixtures

A **heterogeneous mixture** is a **non-uniform** mixture; it is possible to distinguish the component parts from each other, though not always with the naked eye. Heterogeneous mixtures include **suspensions** and **colloids**.

Solutions, suspensions and colloids

Solutions

*A **solution** is a homogeneous mixture of two or more substances; one substance is usually a liquid.*

A solution is composed of:

- The **solvent**, which is the substance that does the **dissolving**. The solvent is present in the higher concentration.
- The **solute**, which is the substance that **dissolves**. The solute is present in the lower concentration. A solution may contain more than one solute.

Solutions in which the solvent is water are known as **aqueous solutions**.

*A **saturated solution** is a solution in which the solvent cannot dissolve any more solute at a particular temperature, in the presence of undissolved solute.*

Table 2.4 *Different types of solutions*

State of solute	State of solvent	Example	Components
Solid	Liquid	Sea water	Sodium chloride dissolved in water
Liquid	Liquid	White vinegar	Ethanoic acid dissolved in water
Gas	Liquid	Soda water	Carbon dioxide dissolved in water
Solid	Solid	Bronze (a metal alloy)	Tin dissolved in copper
Gas	Gas	Air	Oxygen, carbon dioxide, noble gases and water vapour dissolved in nitrogen

Suspensions

*A **suspension** is a heterogeneous mixture in which minute, visible particles of one substance are dispersed in another substance, which is usually a liquid.*

Examples

- **Mud** in **water** and **powdered chalk** in **water**. These are suspensions of solid particles in a liquid.
- **Oil shaken** in **water**. This is a suspension of liquid droplets in a liquid.
- **Dust** in the **air**. This is a suspension of solid particles in a gas.

Colloids

*A **colloid** is a heterogeneous mixture in which minute particles of one substance are dispersed in another substance, which is usually a liquid. The dispersed particles are larger than those of a solution, but smaller than those of a suspension.*

Colloids are **intermediate** between a solution and a suspension.

Table 2.5 *Different types of colloids*

Type of colloid	Composition	Examples
Gel	Solid particles dispersed in a liquid	Gelatin, jelly
Emulsion	Liquid droplets dispersed in a liquid	Mayonnaise, milk, hand cream
Foam	Gas bubbles dispersed in a liquid	Whipped cream, shaving cream
Solid aerosol	Solid particles dispersed in a gas	Smoke
Liquid aerosol	Liquid droplets dispersed in a gas	Fog, aerosol sprays, clouds

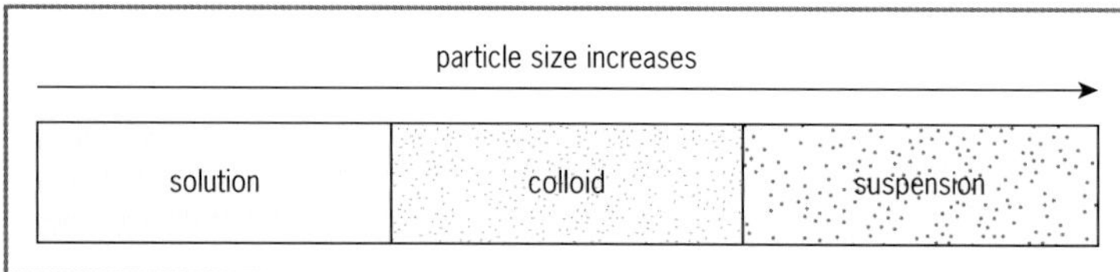

Figure 2.2 *Comparing the particle sizes in a solution, a colloid and a suspension*

Table 2.6 *Comparing the properties of solutions, colloids and suspensions*

Property	Solution	Colloid	Suspension
Size of dispersed particles	Extremely small.	Larger than those in a solution but smaller than those in a suspension.	Larger than those in a colloid.
Visibility of dispersed particles	Not visible, even with a microscope.	Not visible, even with a microscope.	Visible to the naked eye.
Sedimentation	Components do not separate if left undisturbed.	Dispersed particles do not settle if left undisturbed.	Dispersed particles settle if left undisturbed.
Passage of light	Light usually passes through.	Most will scatter light.	Light does not pass through.
Appearance	Usually transparent due to light passing through.	Translucent due to the scattering of light, or may be opaque.	Opaque due to light not being able to pass through.

Solubility

Solubility *is the mass of solute that will saturate 100 g of solvent at a specified temperature.*

In general, the solubility of a **solid** solute in water **increases** as the temperature increases.

Solubility curves

A **solubility curve** is drawn by plotting solubility against temperature, as shown in Figure 2.3.

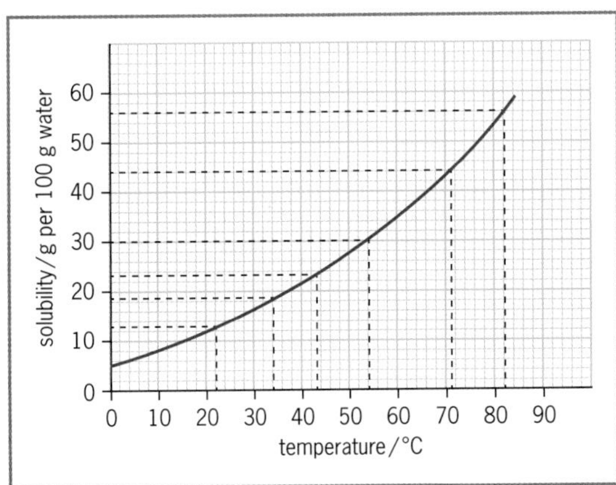

Figure 2.3 *Solubility curve for copper(II) sulfate ($CuSO_4$) in water*

Solubility curves are useful to obtain different pieces of **information**, as shown in the following questions.

Sample questions

1 What is the **solubility** of copper(II) sulfate at 43 °C?

The solubility of $CuSO_4$ at 43 °C = **23 g per 100 g water**

2 A copper(II) sulfate solution containing 100 g water is saturated at 34 °C. What mass of copper(II) sulfate must be added to **re-saturate** this solution if it is heated to 71 °C?

At 34 °C, **19 g** of $CuSO_4$ saturates 100 g of water

At 71 °C, **44 g** of $CuSO_4$ saturates 100 g of water

∴ the mass of $CuSO_4$ to be added to re-saturate a solution containing 100 g of water

= 44 – 19 g

= **35 g**

3 A copper(II) sulfate solution which contains 300 g of water is saturated at 54 °C. What mass of copper(II) sulfate would **crystallise out** of this solution if it is cooled to 22 °C?

At 54 °C, **30 g** of $CuSO_4$ saturates 100 g of water

At 22 °C, **13 g** of $CuSO_4$ saturates 100 g of water

∴ the mass of $CuSO_4$ crystallising out of a saturated solution containing 100 g of water = 30 – 13 g = **17 g**

and mass of $CuSO_4$ crystallising out of a saturated solution containing

$$300 \text{ g of water} = \frac{17}{100} \times 300 \text{ g}$$

$$= \underline{\mathbf{51\ g}}$$

4 At what **temperature** would 84 g of copper(II) sulfate saturate 150 g water?

84 g of $CuSO_4$ saturates 150 g of water

$$\therefore \frac{84}{150} \times 100 \text{ g of } CuSO_4 \text{ saturates 100 g of water}$$

= **56 g of $CuSO_4$**

Temperature at which 56 g of $CuSO_4$ saturates 100 g water = **<u>82 °C</u>**

Separating the components of mixtures

The **technique** used to separate the components of a mixture depends on the **physical properties** of the components.

Filtration

Filtration is used to separate a **suspended** or **settled solid** and a **liquid** when the solid does not dissolve in the liquid, e.g. soil and water. The components are separated due to their different **particle sizes**.

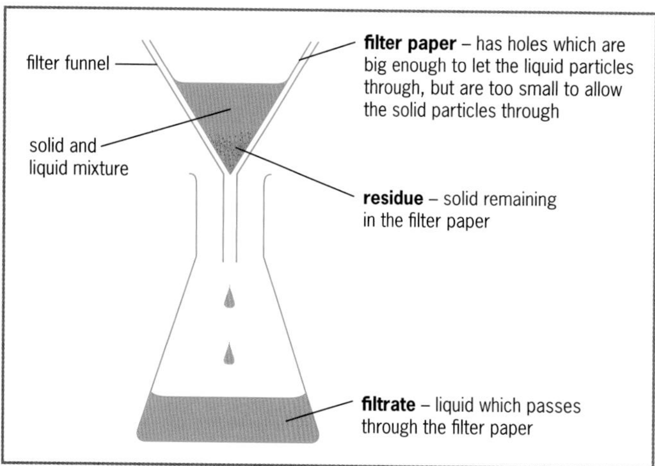

Figure 2.4 *Separating components of a mixture by filtration*

Evaporation

Evaporation is used to separate and retain the **solid solute** from the liquid solvent in a **solution.** It is used if the solute does not decompose on heating or if a solid without water of crystallisation (see p. 72) is required, e.g. to obtain sodium chloride from sodium chloride solution. The components are separated due to their different **boiling points**. The boiling point of the solvent must be lower than that of the solute so that it is converted to a gas and leaves the solute behind.

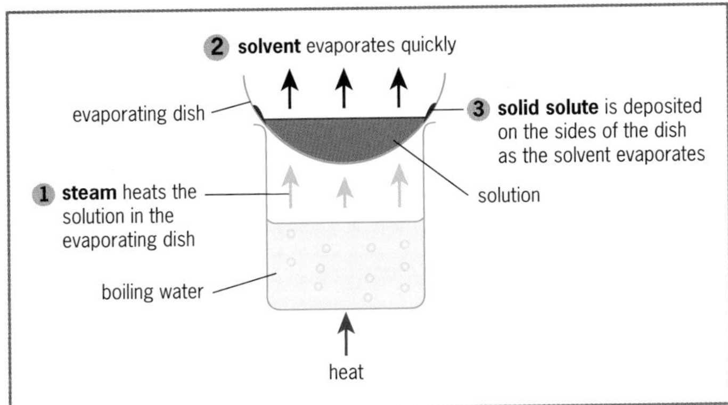

Figure 2.5 *Separating components of a mixture by evaporation*

Crystallisation

Crystallisation is used to separate and retain the **solid solute** from the liquid solvent in a **solution.** It is used if the solute decomposes on heating or if a solid containing water of crystallisation is required, e.g. to obtain hydrated copper(II) sulfate from copper(II) sulfate solution. The components are separated due to their different **volatilities**. The solvent must be more volatile than the solute so that it evaporates and leaves the solute behind.

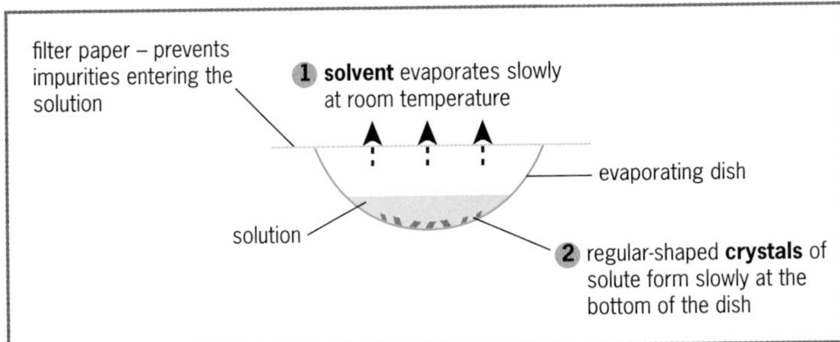

Figure 2.6 *Separating components of a mixture by crystallisation*

Simple distillation

Simple distillation is used separate and retain the **liquid solvent** from the solid solute in a **solution,** e.g. to obtain distilled water from tap water or sea water. The solute can also be obtained by **evaporation** or **crystallisation** of the concentrated solution remaining after distillation if no impurities are present. The components are separated due to their different **boiling points.** The boiling point of the solvent must be lower than that of the solute.

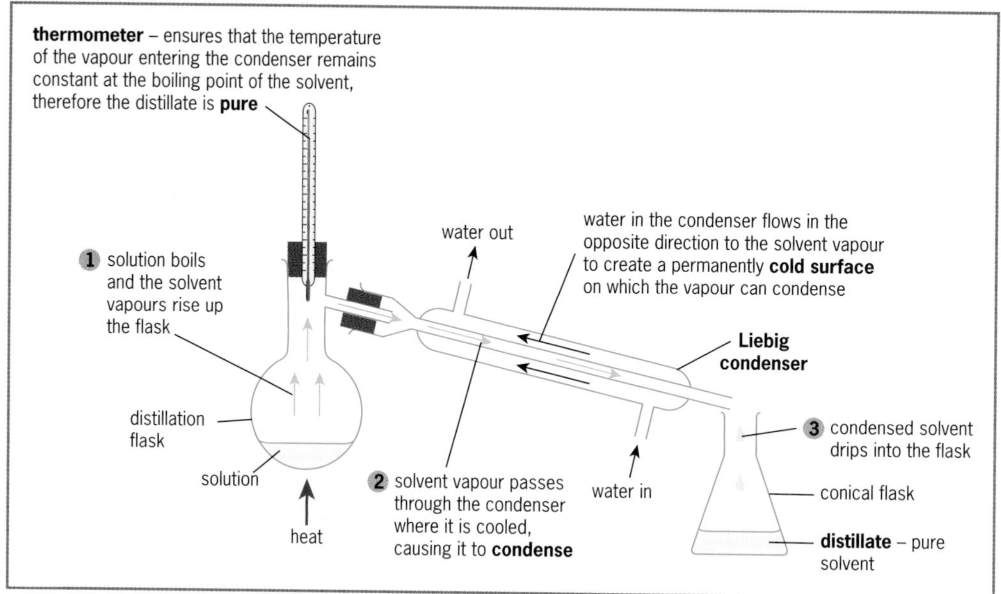

Figure 2.7 *Separating components of a mixture by simple distillation*

Fractional distillation

Fractional distillation is used to separate two (or more) **miscible liquids** with boiling points that are close together, e.g. ethanol, boiling point 78 °C, and water, boiling point 100 °C. Miscible liquids mix completely and are separated due to their different **boiling points**.

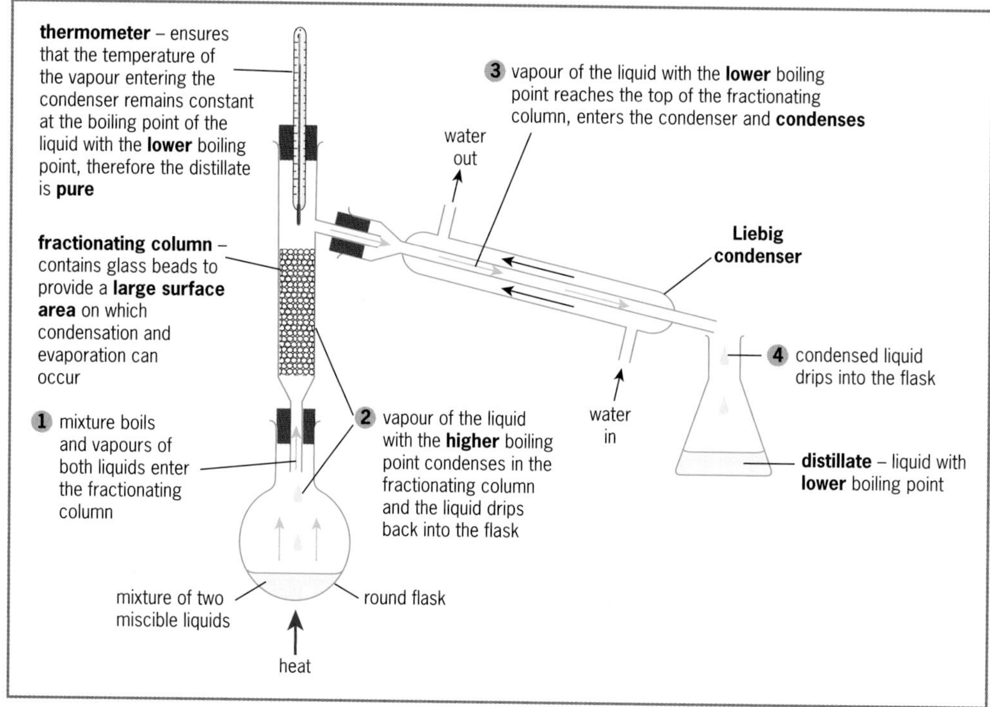

Figure 2.8 *Separating components of a mixture by fractional distillation*

As the mixture boils, vapours of both liquids rise up the fractionating column where they condense and evaporate repeatedly and the vapour mixture becomes progressively richer in the **more volatile** component (the one with the lower boiling point). The vapour reaching the top of the column and entering the condenser is composed almost entirely of the more volatile component and the temperature remains constant at the boiling point of this component.

The temperature begins to rise when almost all of the more volatile liquid has distilled over. This shows that a **mixture** of both liquids is reaching the top of the column and distilling over. This mixture is collected in a second container and discarded. When the temperature reaches the boiling point of the less volatile liquid (the one with the higher boiling point), that liquid is collected in a third container.

Separating funnel

A **separating funnel** is used to separate two (or more) **immiscible liquids**, e.g. oil and water. Immiscible liquids do not mix and are separated due to their **different densities**.

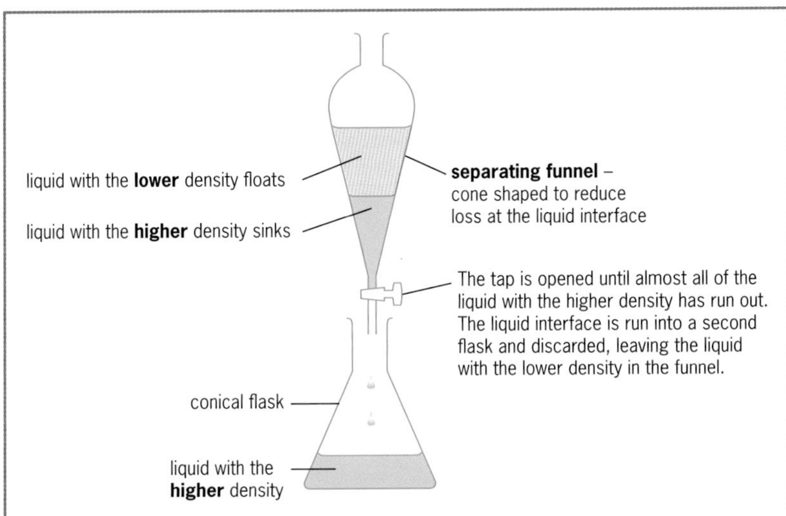

Figure 2.9 *Separating components of a mixture using a separating funnel*

Paper chromatography

Paper chromatography is used to separate **several solutes** which are present in a solution. The solutes are usually coloured and travel through absorbent paper at different speeds, e.g. the dyes in black ink or pigments in chlorophyll. The solutes are separated based on:

- How **soluble** each one is in the solvent used.
- How strongly each one is **attracted** to the paper used.

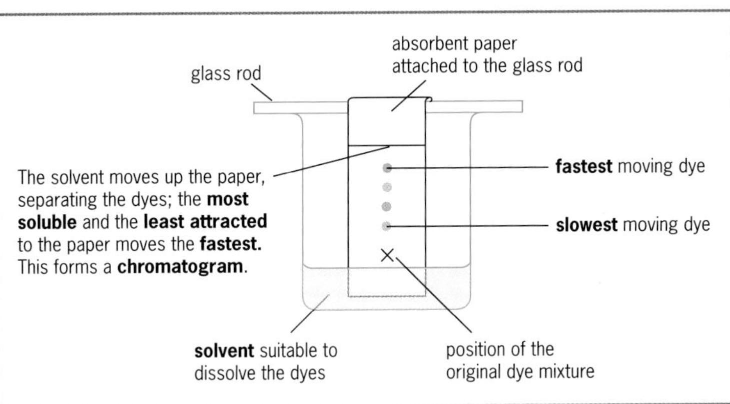

Figure 2.10 *Separating components of a mixture by paper chromatography*

The extraction of sucrose from sugar cane

The extraction of sucrose from sugar cane is an industrial process which uses several **separation techniques.** The extraction process is summarised in Figure 2.11.

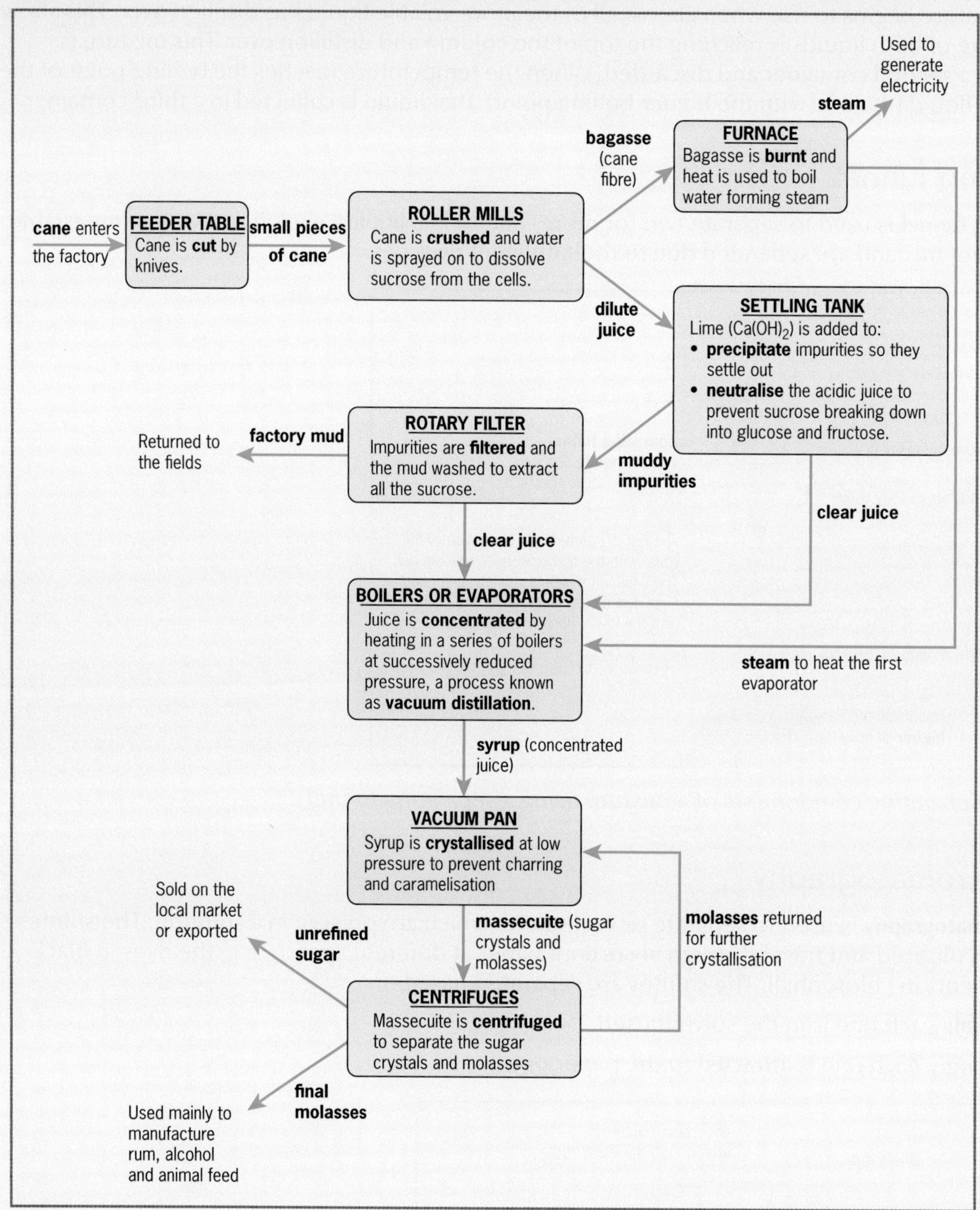

Figure 2.11 *A flow chart showing how sucrose is extracted from sugar cane*

Revision questions

1. Using a table, show THREE differences between a pure substance and a mixture.
2. Define EACH of the following terms:
   - **a** element
   - **b** compound
   - **c** solution
   - **d** suspension
3. Referring to particle size, passage of light and sedimentation, distinguish between a solution, a colloid and a suspension. Give a named example of EACH type of mixture.
4. What is meant by the term 'solubility'?
5. Potassium chlorate(V) ($KClO_3$) was found to have a solubility of 9.0 g per 100 g water at 28 °C and 32.0 g per 100 g water at 74 °C. What mass of potassium chlorate(V) must be added to a solution containing 350 g water which is saturated at 28 °C to make the solution saturated again if it is heated to 74 °C?
6. Draw a labelled diagram of the apparatus you would use to separate sand from water.
7. Explain how you would obtain pure water from tap water. Your answer must include reference to the principles involved.
8. Explain the principles involved in separating each of the following mixtures:
   - **a** cooking oil and water
   - **b** the dyes in a drop of black ink
9. Construct a simple flow diagram to identify the main processes involved in the extraction of sucrose from sugar cane.

3 Atomic structure

Atoms are the basic building blocks of matter.

*An **atom** is the smallest particle of an element that can exist by itself and still have the same chemical properties as the element.*

Subatomic particles

Atoms are made up of **three** different types of fundamental particles called **subatomic particles**:

- **Protons**
- **Neutrons**
- **Electrons**

Protons and neutrons are in a fixed position in the centre, or **nucleus**, of an atom. The electrons are found quite a distance from the nucleus, spinning around the nucleus in **energy shells** (see p. 19).

Table 3.1 *Characteristics of the three subatomic particles*

Particle	Relative charge	Relative mass
Proton	+1	1
Neutron	0	1
Electron	–1	$\frac{1}{1840}$

In any atom, the number of protons and number of electrons is the **same**, therefore atoms have **no overall charge**. Atoms of different elements contain **different numbers** of protons and electrons – this is what makes the elements different. Two numbers can be assigned to any atom, the **atomic number** and the **mass number**.

Atomic number

__Atomic number__ (or proton number) is the number of protons in the nucleus of one atom of an element.

Since the number of electrons in an atom is always equal to the number of protons, the number of **electrons** in an atom is equal to the atomic number. Each element has its own **unique** atomic number.

Mass number

__Mass number__ (or nucleon number) is the total number of protons and neutrons in the nucleus of one atom of an element.

The number of **neutrons** in an atom can be calculated by **subtracting** the atomic number from the mass number. More than one element can have the same mass number, so the number is **not unique** to a particular element.

Nuclear notation

An atom (or ion) of an element can be represented using the **nuclear notation** given below:

$$^{A}_{Z}X$$

where: X = atomic symbol
A = mass number
Z = atomic number

Using the nuclear notation, the number of protons, neutrons and electrons in an atom can be calculated.

Example

The nuclear notation for copper is ${}^{63}_{29}\mathbf{Cu}$. From this, the following can be deduced about a copper atom:

Mass number = 63

Atomic number = 29

Number of protons = 29

Number of neutrons = 63 – 29 = 34

Number of electrons = 29

The nuclear notations for all elements can be found in the **periodic table** of elements on p. 196.

The arrangement of subatomic particles in an atom

Atoms are composed of **two** parts:

- A **nucleus** in the centre containing **nucleons** (**protons** and **neutrons**) packed tightly together. Nearly all the **mass** of an atom is concentrated in its nucleus.
- One or more **energy shells** surrounding the nucleus which contain **electrons** revolving at high speeds. The shells are relatively distant from the nucleus, such that most of an atom is empty space. The electrons moving around the nucleus make up the **volume** of an atom.
  - Each energy shell is a **specific distance** from the nucleus.
  - Electrons with the **least energy** occupy the energy shells closest to the nucleus.
  - Electrons **fill up** the energy shells closest to the nucleus **first**.
  - Each energy shell has a **maximum number** of electrons it can hold:
    - The **first** energy shell can hold up to **two** electrons.
    - The **second** energy shell can hold up to **eight** electrons.
    - The **third** energy shell may be considered to hold up to **eight** electrons.

    Further energy shells can hold more electrons, but it is unnecessary to know their maximum numbers at this level of study.

The arrangement of electrons in an atom is known as its **electronic configuration**. This can be represented in writing using **numbers** or by drawing a **shell diagram**.

Representing atoms

To draw a **shell diagram** of an atom, you first determine the number of protons, neutrons and electrons, and then write its electronic configuration using **numbers** separated by commas.

Examples

The helium atom, ${}^{4}_{2}\mathbf{He}$

A helium atom has:

2 protons

2 neutrons (4 – 2)

2 electrons

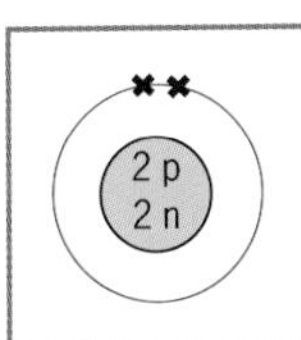

The electronic configuration is **2**

∴ it has **2** electrons in the first shell

The boron atom, $^{11}_{5}B$

A boron atom has:

5 protons

6 neutrons (11 – 5)

5 electrons

The electronic configuration is **2,3**

∴ it has: **2** electrons in the first shell

3 electrons in the second shell

The sulfur atom, $^{32}_{16}S$

A sulfur atom has:

16 protons

16 neutrons (32 – 16)

16 electrons

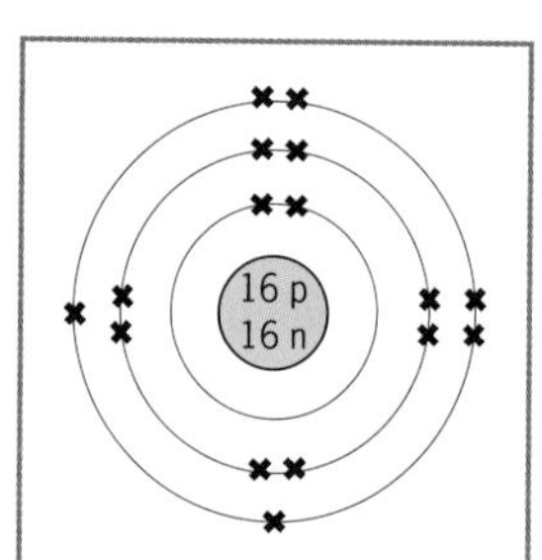

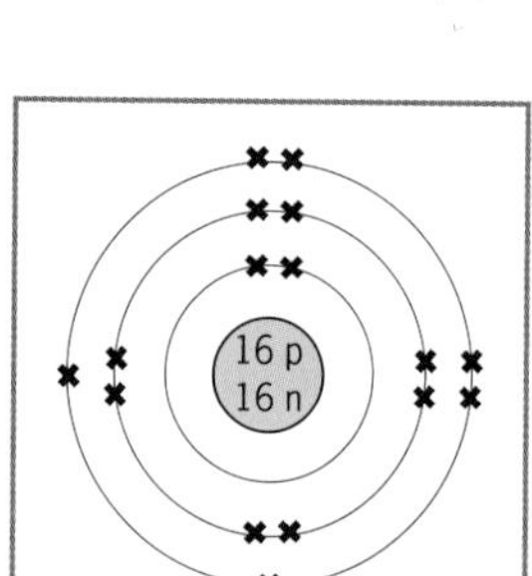

The electronic configuration is **2,8,6**

∴ it has: **2** electrons in the first shell

8 electrons in the second shell

6 electrons in the third shell

Note Electrons are only shown as being paired in the second and third shells when the shells contain five or more electrons.

The **number** and **arrangement** of electrons in an element determine its **chemical properties**, especially the number of electrons in the outermost energy shell which are known as the **valence electrons**.

Isotopes

Isotopes *are different atoms of a single element that have the same number of protons in their nuclei, but different numbers of neutrons.*

Therefore, isotopes of an element have the **same** atomic number but **different** mass numbers.

- Isotopes of an element have the **same chemical properties** because the number and arrangement of electrons in them are the same.
- Isotopes of an element have slightly **different physical properties** because they have different numbers of neutrons which give them slightly different masses.

Isotopy *is the occurrence of atoms of a single element that have the same number of protons in their nuclei, but different numbers of neutrons.*

Examples

Chlorine

Chlorine has **two** naturally occurring isotopes: $^{35}_{17}Cl$ and $^{37}_{17}Cl$.

Table 3.2 *The isotopes of chlorine*

Isotope	Percentage of isotope	Mass number	Atomic number	Number of		
				Protons	Electrons	Neutrons
$^{35}_{17}Cl$	75%	35	17	17	17	**18**
$^{37}_{17}Cl$	25%	37	17	17	17	**20**

Average mass number of naturally occurring chlorine $= \left[\frac{75}{100} \times 35\right] + \left[\frac{25}{100} \times 37\right]$

$= \mathbf{35.5}$

Carbon

Carbon has **three** naturally occurring isotopes: ${}^{12}_{6}C$, ${}^{13}_{6}C$ and ${}^{14}_{6}C$.

Table 3.3 *The isotopes of carbon*

Isotope	Percentage of isotope	Mass number	Atomic number	Number of		
				Protons	Electrons	Neutrons
${}^{12}_{6}C$	98.89%	12	6	6	6	**6**
${}^{13}_{6}C$	1.10%	13	6	6	6	**7**
${}^{14}_{6}C$	0.01%	14	6	6	6	**8**

Carbon-14 is **radioactive.**

Radioactive isotopes

The nuclei of some isotopes are **unstable** and spontaneously undergo **radioactive decay**, during which they eject small particles and **radiation.** These are called **radioactive isotopes** and they eject these particles to become more **stable.** They may produce atoms of one or more different elements at the same time. The **time** taken for half of the nuclei in a sample of a radioactive isotope to undergo radioactive decay is called its **half-life.**

Uses of radioactive isotopes

- **Carbon-14 dating**

  The **age** of plant and animal remains, up to about 60 000 years old, can be determined by **carbon-14 dating.** Living organisms constantly take in carbon from carbon dioxide or food molecules, 0.01% of which is radioactive **carbon-14**. This keeps the percentage of carbon-14 in living organisms constant. When an organism dies, it stops taking in carbon and the percentage of carbon-14 in its body decreases as it undergoes radioactive decay. As the half-life of carbon-14 is **5700 years,** if the percentage of radioactive carbon-14 left in plant and animal remains is measured, it can be used to determine their ages.

- **Cancer treatment (radiotherapy)**

  Cancerous cells in tumours can be destroyed by directing a controlled beam of radiation from radioactive **cobalt-60** at the cells. Alternatively, a radioactive isotope can be injected directly into the cancerous tumour, e.g. radioactive **iodine-131** is used to treat thyroid cancer.

- **Energy generation**

  Electricity is generated in nuclear power stations using radioactive **uranium-235**. If a uranium-235 atom is struck by a fast moving neutron, it splits into two smaller atoms. As it splits, two or three neutrons and a large amount of heat energy are released. The neutrons can then strike other atoms causing them to split and release more neutrons and energy. This causes a **chain reaction** which releases very large amounts of **heat energy**.

  If the chain reaction is controlled, the energy can be used to **generate electricity**. If the chain reaction is not controlled, a **nuclear explosion** (for example, from an atom bomb) can occur.

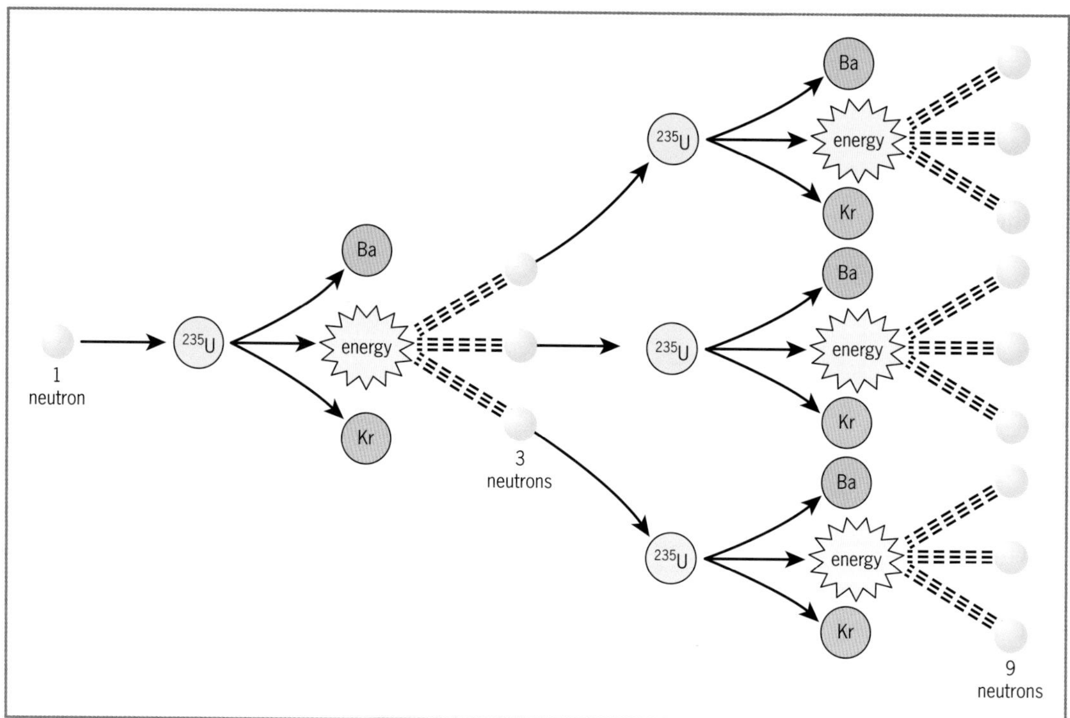

Figure 3.1 *The chain reaction created by splitting a uranium-235 atom*

- **Tracers**

  Very tiny quantities of radioactive isotopes can be observed and traced using special equipment in **medical investigations** and **biological research**. For example, the functioning of the thyroid gland can be checked by giving patients radioactive **iodine-131** and radioactive **carbon-14** is used to study carbon dioxide uptake and photosynthesis in plants.

- **Heart pacemakers**

  Heart pacemakers are usually powered by chemical batteries which have to be replaced by a surgical procedure about every 10 years. Batteries containing **plutonium-238** should be able to power pacemakers for a patient's lifetime without having to be replaced, since the half-life of plutonium-238 is about 87 years.

The mass of atoms

Because an atom of any element is so small, its **absolute mass (actual mass)** is very difficult to measure, e.g. the absolute mass of a hydrogen atom is 1.67×10^{-24} g. Consequently, the masses of atoms are usually **compared** using **relative atomic mass**.

***Relative atomic mass (A_r)** is the average mass of one atom of an element compared to one-twelfth the mass of an atom of carbon-12.*

A carbon-12 atom has been assigned a mass of **12.00 atomic mass units** or **amu**. This means that 1/12th the mass of a carbon-12 atom has a mass of **1.00 amu**, and relative atomic mass **compares** the masses of atoms to this value. Being a comparative value, relative atomic mass has **no units**.

When the relative atomic mass of an element is calculated, the **relative abundance** of each **isotope** is taken into account. As a result, relative atomic masses of elements are not usually whole numbers, e.g. the relative atomic mass of chlorine is 35.5 (see p. 20).

Revision questions

1. What is an atom?

2. Atoms are composed of THREE subatomic particles. Name these and use their relative masses and charges to distinguish between them.

3. Define the following terms:

   a atomic number

   b mass number

4. Draw a shell diagram to show the structure of EACH of the following atoms and identify the atom in EACH case:

   a $^{12}_{6}C$

   b $^{39}_{19}K$

   c $^{35}_{17}Cl$

   d $^{9}_{4}Be$

5. What does the term 'isotopy' mean?

6. Naturally occurring boron consists of 20% $^{10}_{5}B$ and 80% $^{11}_{5}B$.

   a What can you deduce about naturally occurring boron?

   b Determine the average mass number of naturally occurring boron.

7. What is a radioactive isotope?

8. Outline how radioactive isotopes are used to:

   a generate energy in nuclear power stations

   b determine the age of a fossil

   c treat cancer

9. What is relative atomic mass and why is it used to determine the mass of atoms?

4 The periodic table and periodicity

The periodic table is a **classification** of all elements. Elements in the periodic table show **periodicity**.

***Periodicity** is the recurrence of similar chemical and physical properties at regular intervals that is seen in the elements in the periodic table.*

The historical development of the periodic table

Scientists started attempting to **classify** elements early in the nineteenth century.

- **Johann Döbereiner**

  Between 1817 and 1829, Johann Döbereiner proposed the **Law of Triads**. He noticed that certain groups of **three** elements, which he called **triads**, showed similar chemical and physical properties. If the elements in any triad were arranged in increasing relative atomic mass, the relative atomic mass of the middle element was close to the average of the first and third elements. For example, lithium, sodium and potassium have relative atomic masses of 7, 23 and 39.

- **John Newlands**

  In 1865, John Newlands proposed the **Law of Octaves**. He arranged the elements that had been discovered at the time in order of increasing relative atomic mass and found that each element exhibited similar chemical and physical properties to the element **eight** places ahead of it in the list. For example, sodium was eight places ahead of lithium and the two exhibited similar properties. He then placed the similar elements into vertical columns called groups.

- **Dmitri Mendeleev**

  In 1869, Dmitri Mendeleev published his **Periodic Classification of Elements** in which he:

  - Arranged elements in **increasing relative atomic mass**.
  - Placed elements with similar chemical and physical properties together in **vertical** columns (groups).
  - Left **gaps** when it appeared that elements had not yet been discovered.
  - Occasionally ignored the order suggested by relative atomic mass and **exchanged** adjacent elements so they were better classified into chemical families.

  Mendeleev is credited with creating the first version of the periodic table.

- **Henry Moseley**

  In 1914, Henry Moseley placed the elements in increasing **atomic number** which resulted in all elements with similar properties falling in the same groups.

The modern periodic table

The modern periodic table is composed of vertical columns of elements called **groups**, and horizontal rows of elements called **periods**. The elements are organised on the basis of:

- **Increasing atomic number**.
- The **electronic configuration** of their atoms.
- Their **chemical properties**.

Metals are found on the left side of the table and **non-metals** are found on the right side of the table (see Figure 4.1).

Groups

Groups are **vertical columns** of elements. There are 18 groups, eight of which are numbered using Roman numerals from **I** to **VII**, and the last group is Group **0**.

- All elements in the same group have the **same number** of **valence electrons** (electrons in their outermost electron shell).
- For elements in Groups I to VII, the **number of valence electrons** is the same as the **group number**.
- All elements in Group **0** have a **full** outer electron shell.
- Moving **down** any group, each element has **one more electron shell** than the element directly above it.
- All elements in the same group have **similar chemical properties**.
- Moving **down** a group, the **metallic nature** of elements **increases** and the **non-metallic nature decreases.**

Between Groups II and III there are ten groups of elements called the **transition elements** or **transition metals**. Transition metals usually have **two** valence electrons.

Periods

Periods are **horizontal rows** of elements. There are 7 periods, numbered using Arabic numerals from **1** to 7.

- All the elements in the same period have the **same number** of **occupied electron shells**, therefore they have their valence electrons in the same shell.
- The **number of occupied electron shells** is the same as the **period number**.
- Moving **along** any period from left to right, each element has **one more valence electron** than the element directly before it.
- Moving **along** any period from left to right, the **metallic nature** of the elements **decreases** and the **non-metallic nature increases.**

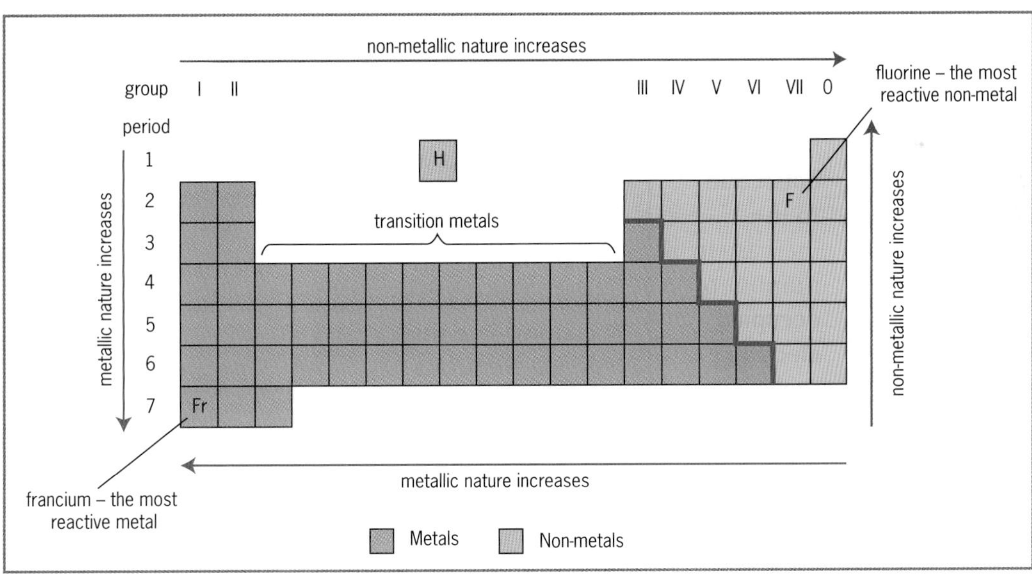

Figure 4.1 *The periodic table in outline*

The periodic table and the electronic configuration of atoms

If the **group number** and **period number** of an atom are known, its electronic configuration can be determined. And if the **electronic configuration** of an atom is known, the group number and period number can be determined using the following:

- The **group number** and the **number of valence electrons** are the same.
- The **period number** and the **number of occupied electron shells** are the same.

Examples

Potassium

Potassium is in **Group I** and **Period 4**.

A potassium atom has: **1** valence electron
4 occupied electron shells

Therefore the **electronic configuration** of a potassium atom is **2,8,8,1**

Silicon

The **electronic configuration** of a silicon atom is **2,8,4**

A silicon atom has: **4** valence electrons
3 occupied electron shells

Therefore silicon is **Group IV** and **Period 3**

Trends in Group II – the alkaline earth metals

Elements in Group II all have similar chemical properties because their atoms all have **two** valence electrons. They react by **losing** these valence electrons to form positively charged ions called **cations** (see p. 32). When they lose these electrons they are said to **ionise**. The easier an element ionises, the more reactive it is. The **ease of ionisation** increases moving **down** Group II, therefore the **reactivity** of the elements increases moving **down** the group.

Table 4.1 *Summary of the trends in Group II – the alkaline earth metals*

Element	Number of occupied electron shells	Atomic radius	Ease of ionisation	Reactivity with oxygen, water and dilute hydrochloric acid
Be	2	↓ Increases moving **down** due to the increase in number of **occupied electron shells.**	↓ Increases moving **down**. As the atomic radii **increase**, the attractive pull of the positive nucleus on the valence electrons decreases, and the more easily the atoms **lose** their valence electrons to form cations.	↓ Increases moving **down** due to the increase in the **ease of ionisation.**
Mg	3			
Ca	4			
Sr	5			
Ba	6			
Ra	7			

⟶ Indicates that the property **increases** in the direction of the arrow

Reactions of Group II elements

Examples of reactions of calcium

With oxygen:

$$2Ca(s) + O_2(g) \longrightarrow 2CaO(s)$$

With water:

$$Ca(s) + 2H_2O(l) \longrightarrow Ca(OH)_2(aq) + H_2(g)$$

With hydrochloric acid:

$$Ca(s) + 2HCl(aq) \longrightarrow CaCl_2(aq) + H_2(g)$$

Trends in Group VII – the halogens

Elements in Group VII exist as **diatomic molecules**, these being F_2, Cl_2, Br_2 and I_2. They all have similar chemical properties because their atoms all have **seven** valence electrons. They react by **gaining** one valence electron to form negatively charged ions called **anions** (see p. 32). When they gain this electron they are said to **ionise**. The **ease of ionisation** increases moving **up** Group VII, therefore the **reactivity** of the elements increases moving **up** the group.

Table 4.2 *Summary of the trends in Group VII – the halogens*

Element	Appearance and state at room temperature	Number of occupied electron shells	Atomic radius	Ease of ionisation	Reactivity	Strength of oxidising power	Displacement
F	Pale yellow gas	2	↓ Increases moving **down** due to the increase in number of **occupied electron shells**.	↑ Increases moving **up**. As the atomic radii **decrease**, the attractive pull of the positive nucleus on the electron to be gained increases and the more easily the atoms **gain** this electron to form anions.	↑ Increases moving **up** due to the increase in **ease of ionisation**.	↑ Increases moving **up** due to the increase in **ease of ionisation**. The more easily the element ionises, the more easily it **takes** electrons from another reactant.	An element is displaced from its compounds by an element **above** it in the group.
Cl	Yellow-green gas	3					
Br	Red-brown liquid	4					
I	Grey-black solid	5					

→ Indicates that the property **increases** in the direction of the arrow

Displacement reactions and strength of oxidising power

In a **displacement reaction** an element in its free state takes the place of another element in a compound. A **more reactive** element will displace a **less reactive** element. **Chlorine** will displace bromine and iodine, and **bromine** will displace iodine from their compounds.

e.g. $Cl_2(g) + 2KBr(aq) \longrightarrow 2KCl(aq) + Br_2(aq)$

Displacement reactions can be explained by looking at the relative **strength of oxidising power** of the elements. This is determined by how easily one substance **takes electrons** from another substance. The strength of oxidising power of Group VII elements increases moving **up** the group because the ability to **ionise** and **take electrons** from another reactant increases moving upwards. **Chlorine** will take electrons from bromide (Br^-) and iodide (I^-) ions and **bromine** will take electrons from iodide (I^-) ions:

e.g. $Cl_2(g) + 2Br^-(aq) \longrightarrow 2Cl^-(aq) + Br_2(aq)$

Trends in Period 3

Moving along Period 3 from left to right, the **metallic** nature of the elements **decreases** and the **non-metallic** nature **increases.** Silicon in Group IV is a **semi-metal** or **metalloid.** Each element has **three** occupied electron shells.

- The **ease of ionisation** and **reactivity** of the **metals** sodium, magnesium and aluminium **decreases** moving along the period.
- The **ease of ionisation** and **reactivity** of the **non-metals** phosphorus, sulfur and chlorine **increases** moving along the period.
- **Silicon** does not usually ionise, it usually reacts by sharing electrons with other non-metal atoms.
- **Argon** does not ionise and is chemically unreactive.

Table 4.3 *Summary of the trends in Period 3*

Element	Na	Mg	Al	Si	P	S	Cl	Ar
Electronic configuration	2,8,1	2,8,2	2,8,3	2,8,4	2,8,5	2,8,6	2,8,7	2,8,8
Metal/ non-metal	Metal	Metal	Metal	Semi-metal	Non-metal	Non-metal	Non-metal	Non-metal
Electrical conductivity	Good conductors			Semi-conductor	Non-conductors (insulators)			
Electrons lost, gained or shared	Loses $1e^-$	Loses $2e^-$	Loses $3e^-$	Shares $4e^-$	Gains $3e^-$	Gains $2e^-$	Gains $1e^-$	None
Atomic radius	**Decreases** moving from **left to right** due to the increase in number of **positive protons** causing the attractive pull of the positive nucleus on the valence electrons to get stronger							
Ease of ionisation	← Increases moving from **right to left**. As the atomic radii **increase** and the number of positive protons decreases, the more easily the atoms **lose** electrons to form positive cations.			Does not usually ionise	→ Increases moving from **left to right**. As the atomic radii **decrease** and the number of positive protons increases, the more easily the atoms **gain** electrons to form negative anions.			Does not ionise
Reactivity	← Increases moving from **right to left** due to the increase in **ease of ionisation.**				→ Increases moving from **left to right** due to the increase in **ease of ionisation.**			Unreactive

→ Indicates that the property **increases** in the direction of the arrow

Revision questions

1. Outline the contributions of EACH of the following scientists to the development of the periodic table:

   a Johann Döbereiner

   b John Newlands

   c Dmitri Mendeleev

2. What are the features on which the arrangement of elements in the modern periodic table is based?

3. **a** What do all the elements in the same group of the periodic table have in common?

   b What do all the elements in the same period of the periodic table have in common?

4. The electronic configuration of an atom of element X is 2,8,5. Give the group number and period number of element X.

5. Which element, magnesium or calcium, would you expect to react more vigorously with dilute hydrochloric acid? Explain your answer.

6. How does the state of the elements in Group VII at room temperature change moving down the group?

7. Would you expect a reaction to occur if chlorine gas is bubbled into a potassium bromide solution? Explain your answer based on the relative strength of oxidising power of chlorine and bromine.

8. How does the metallic nature of elements change moving from left to right across Period 3?

9. Which element, sulfur or chlorine, would you expect to be more reactive? Explain your answer.

5 Structure and bonding

Atoms of elements in **Group 0** of the periodic table are **stable** and **unreactive** and exist in nature as individual atoms because they have **full** outer electron shells or valence shells. Atoms of all other elements are **not stable** because they do not have full valence shells. These atoms attempt to obtain full valence shells and become **stable** by **bonding** with each other. They can do this by:

- **Losing valence electrons** to atoms of another element. **Metal** atoms with 1, 2 or 3 valence electrons usually lose their valence electrons and form positive **cations.**
- **Gaining valence electrons** from atoms of another element. **Non-metal** atoms with 5, 6 or 7 valence electrons usually gain electrons into their valence shell and form negative **anions.**
- **Sharing electrons** in their valence shells with other atoms. When **non-metal** atoms with 4, 5, 6 or 7 valence electrons bond with each other, they share valence electrons and form **molecules.**

There are three main types of **chemical bonding:**

- **Ionic bonding.** This occurs when a metal bonds with a non-metal (see p. 32).
- **Covalent bonding.** This occurs when two or more non-metals bond (see p. 36).
- **Metallic bonding.** This occurs within metals (see p. 39).

Chemical compounds are formed when elements bond by ionic or covalent bonding.

Chemical formulae

Chemical formulae can be used to represent **compounds** formed by ionic or covalent bonding. A chemical formula shows which **elements** are present in a compound and shows the **ratio** between the elements. Chemical formulae can be written in three main ways:

- The **molecular formula.** This gives the actual number of atoms of each element present in one molecule of a compound. For example, the molecular formula of ethene is C_2H_4
- The **structural formula.** This is a diagrammatic representation of one molecule of the compound. Lines between the atoms are used to represent bonds. For example, the structural formula of ethene is:

```
H       H
 \     /
  C = C
 /     \
H       H
```

- The **empirical formula.** This gives the simplest whole number ratio between the elements in the compound. For example, the empirical formula of ethene is CH_2

How to write empirical formulae of compounds formed from two elements

Empirical formulae of compounds formed from two different elements can be written using the concept of **valence number** or **valency.** Valency is the number of bonds an atom can form when bonding with other atoms. It is determined by the number of valence electrons an atom has, and it can be thought of as the number of electrons an atom has to lose, gain or share when bonding.

Table 5.1 *Valence numbers or valency*

Group number	I	II	Transition metals	III	IV	V	VI	VII	0
Valency	1	2	Variable, often 2	3	4	3	2	1	0

- Elements in Groups **I** to **IV**: valency = the group number
- Elements in Groups **V** to **VII**: valency = 8 – the group number

When forming a compound, both types of atom must lose, gain or share the **same number** of valence electrons. Consequently, the **sum** of the valencies of each element in the compound must be equal.

Example

The empirical formula of calcium nitride is $\mathbf{Ca_3N_2}$:

- Number of atoms of each element from the formula: Ca = **3**
  N = **2**
- Valency of each element: Ca = 2 (Group II)
  N = 3 (Group V)
- Sum of the valencies: Ca = $\mathbf{3} \times 2 = 6$
  N = $\mathbf{2} \times 3 = 6$

This shows the **sum** of the valencies of the two elements is equal, here they are both **6**.

To **write the empirical formula** of a compound formed from two elements:

- Determine the valencies of each element in the compound.
- Write the **symbol** of the first element. If a metal is present, always write its symbol first.
- Write the **valency** of the second element immediately after the symbol of the first element in **subscript**.
- Write the **symbol** of the second element immediately after the subscript.
- Write the **valency** of the first element immediately after the symbol of the second element in **subscript**.

Note If a valency is **1**, then no number is written as a subscript (see **point 1** below).

Example

Aluminium sulfide

- Valency of each element: Al = **3** (Group III)
  S = **2** (Group VI)
- Symbol of the first element followed by the valency of the second element in subscript: $\mathbf{Al_2}$
- Symbol of the second element followed by the valency of the first element in subscript: $\mathbf{Al_2S_3}$

The empirical formula of aluminium sulfide is $\mathbf{Al_2S_3}$

Points to note:

1 Some transition metals have **more than one** valency. The valency is indicated by a Roman numeral placed in brackets after the **name** of the metal, for example:

Copper(II) chloride

Valency of each element: Cu = **2** (given in the name)
Cl = **1** (Group VII)

Empirical formula of copper(II) chloride is $\mathbf{CuCl_2}$

2 If the ratio of the subscripts is not in its simplest form, then cancel to its **simplest form**, for example:

Silicon dioxide

Valency of each element: Si = **4** (Group IV)
O = **2** (Group VI)

Empirical formula of silicon dioxide is $\mathbf{SiO_2}$ (Si_2O_4 is cancelled to its simplest ratio)

Ionic bonding

Ionic bonding occurs between a **metal** and a **non-metal**. Valence electrons are **transferred** from the metal atom (or atoms) to the non-metal atom (or atoms). The metal atoms form **positive ions** known as **cations**. The non-metal atoms form **negative ions** known as **anions**. Both types of ion have full outer electron shells and are stable.

Examples

Sodium chloride   Formula: **NaCl**

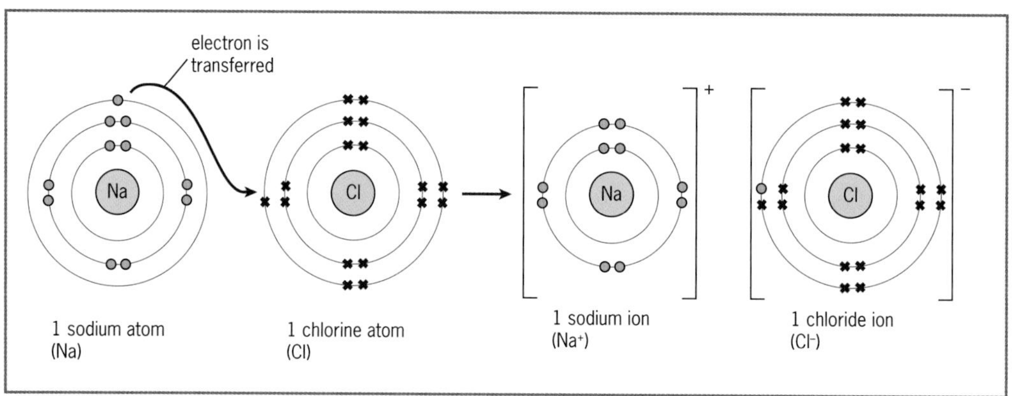

The sodium atom transfers its valence electron to the chlorine atom. A **sodium ion (Na^+)**, which has a single **positive** charge, and a **chloride ion (Cl^-)**, which has a single **negative** charge, are formed.

Lithium oxide   Formula: **Li_2O**

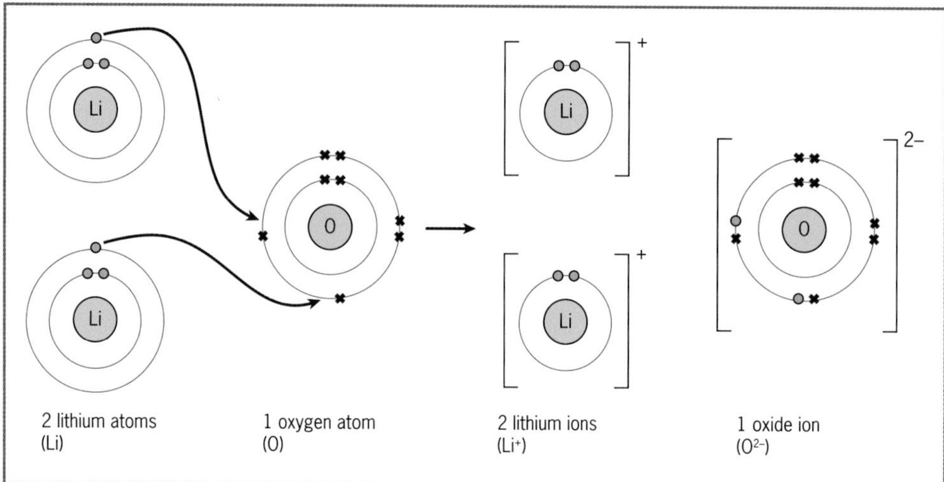

Note Diagrams to show the formation of ionic compounds can be simplified by showing only the **valence electrons**.

Aluminium fluoride Formula: AlF_3

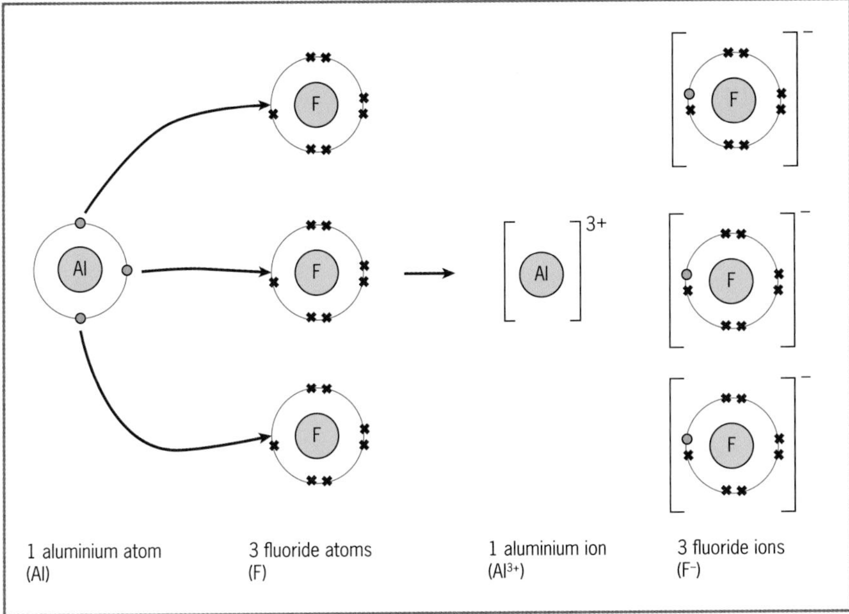

The crystalline structure of ionic compounds

At room temperature, ionic compounds exist as **crystalline solids**. Strong **electrostatic forces** of attraction between the ions, called **ionic bonds**, hold the oppositely charged ions together in a regular, repeating, three-dimensional arrangement throughout the crystal. This forms a structure known as a **crystal lattice**.

Example

Sodium chloride

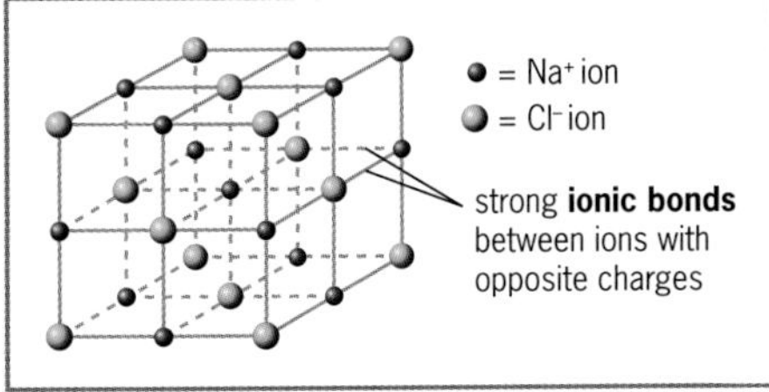

Figure 5.1 *The sodium chloride crystal lattice*

Each sodium ion, Na^+, is bonded to six chloride ions, Cl^-. Each Cl^- ion is bonded to six Na^+ ions. This is seen by looking at the Na^+ ion in the centre of the cube.

Formulae and names of ionic compounds

Ionic compounds can be composed of ions formed from **single atoms** called **monatomic ions** (see p. 32), or ions formed from small **groups of atoms** which are bonded together and called **polyatomic ions**, e.g. the ammonium ion, NH_4^+, and the carbonate ion, CO_3^{2-}.

Table 5.2 *Common cations*

Monovalent	
hydrogen	H^+
lithium	Li^+
sodium	Na^+
potassium	K^+
copper(I)	Cu^+
silver	Ag^+
ammonium	NH_4^+

Divalent	
magnesium	Mg^{2+}
calcium	Ca^{2+}
barium	Ba^{2+}
iron(II)	Fe^{2+}
copper(II)	Cu^{2+}
zinc	Zn^{2+}
tin(II)	Sn^{2+}
lead(II)	Pb^{2+}

Trivalent	
iron(III)	Fe^{3+}
aluminium	Al^{3+}

Table 5.3 *Common anions*

Monovalent	
fluoride	F^-
chloride	Cl^-
bromide	Br^-
iodide	I^-
hydride	H^-
hydroxide	OH^-
nitrite (nitrate(III))	NO_2^-
nitrate (nitrate(V))	NO_3^-
hydrogen carbonate	HCO_3^-
hydrogen sulfate	HSO_4^-
manganate(VII)	MnO_4^-
ethanoate	CH_3COO^-

Divalent	
oxide	O^{2-}
sulfide	S^{2-}
sulfite (sulfate(IV))	SO_3^{2-}
sulfate (sulfate(VI))	SO_4^{2-}
carbonate	CO_3^{2-}
dichromate(VI)	$Cr_2O_7^{2-}$

Trivalent	
nitride	N^{3-}
phosphate	PO_4^{3-}

To **name anions:**

- The name of an anion formed from a **single atom** is derived from the name of the element, with the ending '**-ide**'. For example, N^{3-} is the nitr**ide** ion.
- When **oxygen** is present in a **polyatomic anion** the name of the ion is derived from the element combined with the oxygen, with the ending '**-ite**' or '**-ate**'. For example, NO_2^- is the nitr**ite** ion and NO_3^- is the nitr**ate** ion. Alternatively, the ending '**-ate**' is used with the **oxidation number** of the element given in brackets. For example, NO_2^- is the nitr**ate(III)** ion and NO_3^- is the nitr**ate(V)** ion (see p. 83).

When **writing formulae** of ionic compounds the **sum** of the **positive charges** and the **sum** of the **negative charges** must be **equal**. This is because the **total number** of electrons lost by one type of atom or group of atoms must be the **same** as the total number gained by the other type of atom or group of atoms. Formulae of ionic compounds are **empirical formulae** since they represent the **ratio of ions** present. They are also known as **formula units**.

Example

The chemical formula of magnesium nitride is $\mathbf{Mg_3N_2}$:

- Number of atoms of each element from the formula: Mg = **3**
  N = **2**
- Charge on each ion: Mg^{2+} = +2
  N^{3-} = −3
- Sum of the charges: Mg = **3** × +2 = +6
  N = **2** × −3 = −6

This shows the **sum** of the positive charges and the **sum** of the negative charges are equal.

To **write the chemical formula** of an ionic compound:

- Write down the formulae of the two ions present from the Tables 5.2 and 5.3.
- Rewrite the **formula** of the cation without its charge.
- Write the **magnitude of the charge** on the anion immediately after the formula of the cation in **subscript**.
- Write the **formula** of the anion immediately after the subscript without its charge.
- Write the **magnitude** of the charge on the cation immediately immediately after the formula of the anion in **subscript**.

Note If the magnitude of the charge is **1**, then no number is written as a subscript (see examples below).

Example

Aluminium oxide

- Ions present: Al^{3+} O^{2-}
- Formula of the cation without its charge: **Al**
- Magnitude of the charge on the anion written after the cation in subscript: Al_2
- Formula of the anion without its charge: Al_2O
- Magnitude of the charge on the cation written after the anion in subscript: Al_2O_3

The empirical formula of aluminium oxide is Al_2O_3

Sodium carbonate

- Ions present: Na^+ CO_3^{2-}
- Magnitude of the charges: $Na^+ = \mathbf{1}$
  $CO_3^{2-} = \mathbf{2}$
- Empirical formula of sodium carbonate is Na_2CO_3

Points to note:

1 If either ion is a **polyatomic ion** and **more than one** is required, place the polyatomic ion in **brackets** and write the required subscript outside the bracket, for example:

Iron(III) nitrate

Ions present: Fe^{3+} NO_3^-

Magnitude of the charges: $Fe^{3+} = \mathbf{3}$

$NO_3^- = \mathbf{1}$

Empirical formula of iron(III) nitrate is $Fe(NO_3)_3$ (NO_3 is placed in brackets with the subscript of 3 outside showing that 3 NO_3^- ions are required)

2 If the ratio of the subscripts is not in its simplest form, then cancel to its **simplest form**, for example:

Calcium sulfate

Ions present: Ca^{2+} SO_4^{2-}

Magnitude of the charges: $Ca^{2+} = \mathbf{2}$

$SO_4^{2-} = \mathbf{2}$

Empirical formula of calcium sulfate is $CaSO_4$ ($Ca_2(SO_4)_2$ is cancelled to its simplest ratio)

Covalent bonding

Covalent bonding occurs when two or more **non-metal** atoms bond. Unpaired valence electrons are **shared** between the atoms which results in the formation of **molecules**. The shared electrons orbit around both atoms sharing them. This forms strong **covalent bonds** which hold the atoms together. Each shared pair of electrons forms one covalent bond.

Covalent bonding can occur between atoms of the **same element**, e.g. fluorine (F_2). It can also occur between atoms of two or more **different elements**, e.g. carbon dioxide (CO_2). **Seven** common elements are composed of **diatomic molecules** in their free state:

hydrogen (H_2), oxygen (O_2), nitrogen (N_2), fluorine (F_2), chlorine (Cl_2), bromine (Br_2) and iodine (I_2).

The formula of a covalent substance represents **one molecule** of the substance, therefore, it is the **molecular formula**. In many covalent compounds, the molecular and empirical formulae are the same, e.g. water (H_2O). In some they are not the same, e.g. the molecular formula of glucose is $C_6H_{12}O_6$ but its empirical formula is CH_2O.

Examples

Fluorine Formula: F_2

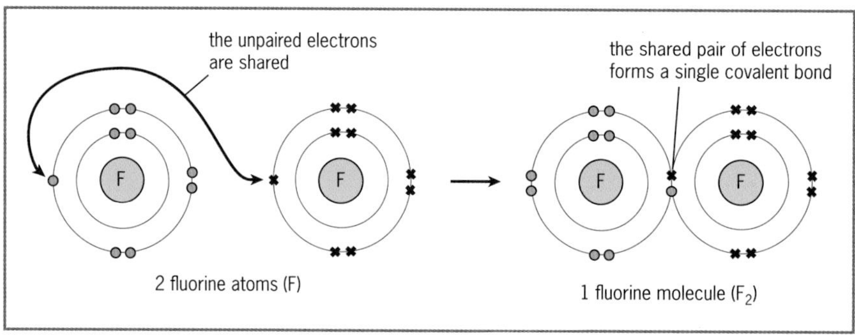

Carbon dioxide Formula: CO_2

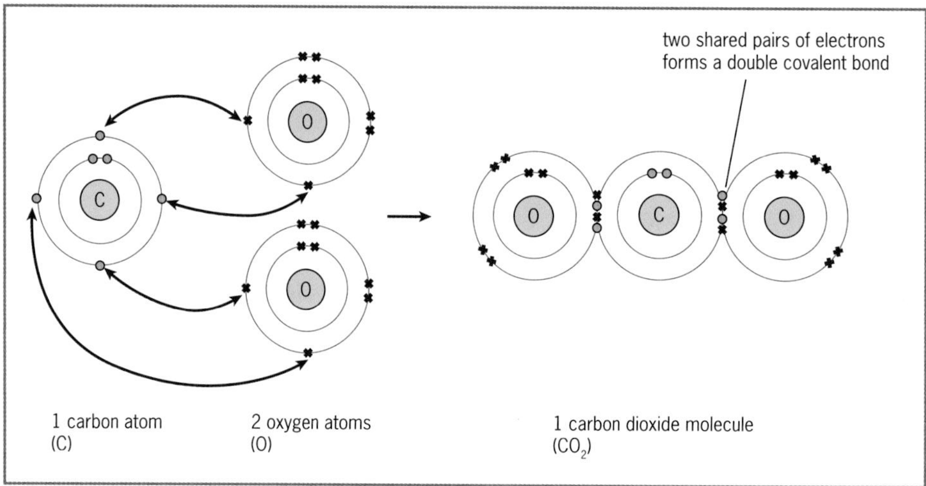

Note Diagrams to show the formation of covalent substances can be simplified by showing only the **valence electrons**.

Nitrogen   Formula: N_2

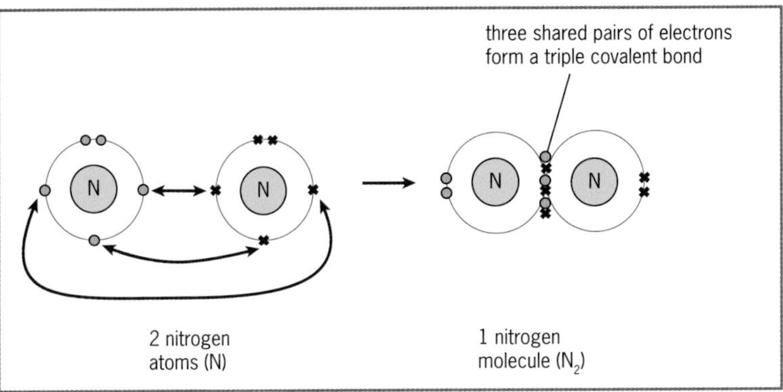

Structural formulae can also be used to represent molecules. Each covalent bond is shown by a line between the two atoms.

F – F   O = C = O   N ≡ N

Figure 5.2 *Structural formulae of fluorine, carbon dioxide and nitrogen*

The structure of covalent substances

Covalent substances are composed of **individual molecules** which can be either **polar** or **non-polar** due to the **electronegativity** of the atoms present. Electronegativity is a measure of how strongly atoms attract bonding electrons. Fluorine, oxygen, chlorine and nitrogen are the most electronegative elements.

- In a **polar molecule**, one type of atom has a **partial positive charge (δ+)** and another type has a **partial negative charge (δ–)** because the atoms at either side of the covalent bond differ in electronegativity and attract the shared electrons with different strengths. Examples are water (H_2O), ammonia (NH_3), hydrogen chloride (HCl) and ethanol (C_2H_5OH).

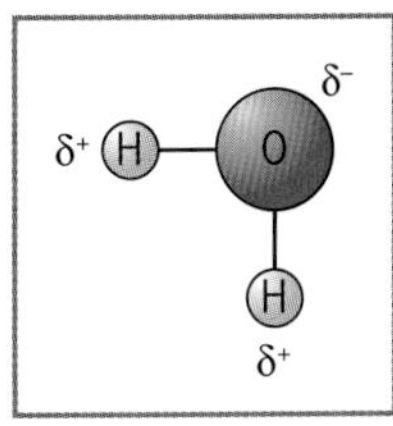

Figure 5.3 *A polar water molecule*

- In a **non-polar molecule**, the electronegativity of the atoms is similar or the same and they attract the shared electrons with equal strengths so the molecule does not have partially charged regions. Examples are hydrogen (H_2), oxygen (O_2) and methane (CH_4).

The **covalent bonds** holding the atoms together in the molecules are **strong**. The molecules themselves are held together by **intermolecular forces** which are **weak** in polar substances and **extremely weak** in non-polar substances.

Drawing dot and cross bonding diagrams

To **draw** dot and cross diagrams to show the formation of ionic and covalent compounds:

- Decide if the compound is **ionic** or **covalent**. If it is formed from a metal and a non-metal it is ionic. If it is formed from two or more non-metals it is covalent.
- Determine the **formula** of the compound using the formulae of the ions, or valency.
- Draw **each atom** in the formula, showing either all the electron shells or just the valence electrons. Use different symbols for electrons of each different type of atom, such as o and x.

- Draw **arrows** to indicate electrons which are transferred or shared.
- Redraw the **ions** formed after electrons have been transferred, or the **molecule** formed after electrons have been shared. Do not forget to put the **charges** on all ions.

Sample questions

Draw a dot and cross diagram to show how each of the following compounds is formed:

a silicon tetrachloride

b magnesium nitride

a Silicon tetrachoride

Type of compound: **covalent** since it is formed from two non-metals

Valencies: Si = 4, Cl = 1

Formula: $\mathbf{SiCl_4}$

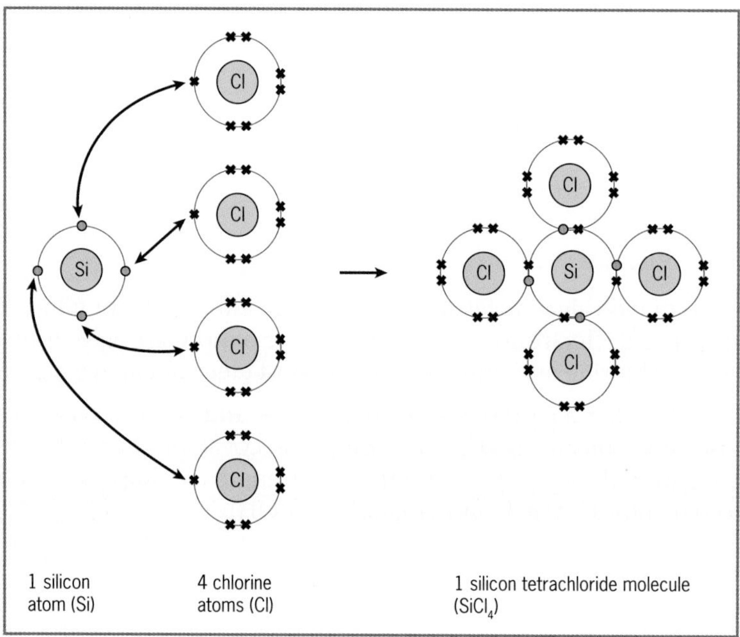

b Magnesium nitride

Type of compound: **ionic** since it is formed from a metal and a non-metal

Ions present: Mg^{2+}, N^{3-}

Formula: $\mathbf{Mg_3N_2}$

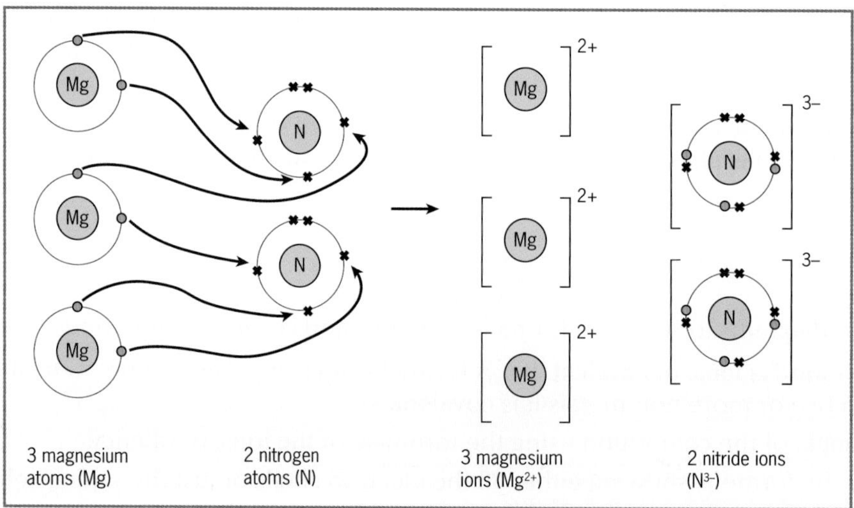

Metallic bonding

Metallic bonding occurs in **metals**. The metal atoms are packed tightly together in rows to form a **metal lattice**, and their valence electrons become **delocalised**. This means that the valence electrons are no longer associated with any specific atom and are free to move. This forms positive **cations** and a **'sea' of mobile electrons**. The metal lattice is held together by the electrostatic forces of attraction between the delocalised electrons and the cations, known as the **metallic bond**, which is strong.

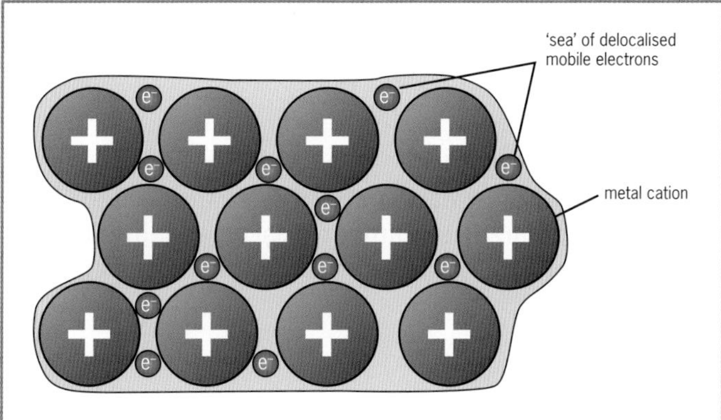

Figure 5.4 *A metal lattice*

Physical properties of metals

The way the atoms in metals are bonded helps to explain their **physical properties**.

Table 5.4 *Physical properties of metals*

Physical property	Explanation
High melting points and boiling points	The strong electrostatic forces of attraction between the cations and delocalised electrons require large amounts of heat energy to break.
Solid at room temperature (except mercury)	Room temperature is not high enough to break the strong electrostatic forces of attraction between the cations and delocalised electrons.
High density	The atoms are packed very closely together.
Conduct electricity	The delocalised electrons are free to move and carry electricity through the metal.
Conduct heat	The delocalised electrons move and carry heat through the metal.
Malleable and ductile	The atoms of each metal are all of the same type and size. If force is applied, the atoms can slide past each other into new positions without the metallic bonds breaking.

The structure and properties of solids

Solids can be divided into **four** groups based on their structure:

- **Ionic** crystals
- **Simple molecular** crystals
- **Giant molecular** crystals
- **Metallic** crystals (see 'Metallic bonding' above)

Ionic crystals

An **ionic crystal** is made of an **ionic lattice** in which strong electrostatic forces of attraction called **ionic bonds** hold the **cations** and **anions** together in a regular, repeating, three-dimensional arrangement. Ionic crystals are represented by **empirical formulae** or **formula units**.

Examples

- **Sodium chloride**, empirical formula **NaCl**. Made of Na^+ ions and Cl^- ions (see p. 33).
- All other **ionic compounds**.

Simple molecular crystals

A **simple molecular crystal** is made of a **molecular lattice** in which weak forces of attraction called **intermolecular forces** hold **small molecules** together in a regular, three-dimensional arrangement. The atoms within each molecule are bonded together by strong covalent bonds. Simple molecular crystals are represented by **molecular formulae**.

Examples

- **Ice**, molecular formula H_2O. Made of water molecules.
- **Dry ice**, molecular formula CO_2. Made of carbon dioxide molecules.
- **Iodine**, molecular formula I_2. Made of iodine molecules.
- **Glucose**, molecular formula $C_6H_{12}O_6$. Made of glucose molecules.

Table 5.5 *Distinguishing between ionic and simple molecular solids*

Property	Ionic solids	Simple molecular solids
Structure	Composed of **ions** held together by strong ionic bonds.	Composed of **molecules** with strong covalent bonds between the atoms in the molecules and weak intermolecular forces between molecules.
Melting point	**High** – the strong ionic bonds between the ions require large amounts of heat energy to break.	**Low** – the weak intermolecular forces between the molecules require little heat energy to break.
Solubility	Most are **soluble** in **water**, a polar solvent, but are insoluble in non-polar organic solvents such as kerosene, gasoline and tetrachloromethane.	Most are **soluble** in **non-polar organic solvents**, but are insoluble in water. Polar compounds are soluble in water, e.g. glucose.
Conductivity	**Do not** conduct electricity when **solid** – the ions are held together by strong ionic bonds and are not free to move. **Do** conduct electricity when **molten** (melted) or **dissolved in water** – the ionic bonds have broken and the ions are free to move and carry electricity.	**Do not** conduct electricity in any state – they do not have any charged particles which are free to move.

Giant molecular crystals

A **giant molecular crystal** is composed of a **giant molecular lattice** in which strong **covalent bonds** hold non-metal atoms together in a regular, three-dimensional arrangement throughout the lattice. Giant molecular crystals are represented by **empirical formulae**.

Examples

- **Diamond**, empirical formula **C**. Made of carbon atoms.
- **Graphite**, empirical formula **C**. Made of carbon atoms.
- **Silicon dioxide**, empirical formula **SiO_2**. Made of silicon and oxygen atoms.

Diamond and graphite are known as **allotropes** of carbon.

__Allotropes__ are different structural forms of a single element in the same physical state.

- Allotropes have the same **chemical properties** because they are both composed of the **same** element.
- Allotropes have different **physical properties** because the atoms are **bonded differently**.

__Allotropy__ is the existence of different structural forms of a single element in the same physical state.

Diamond

In **diamond**, each carbon atom is bonded covalently to **four** others, which are arranged in a **tetrahedron** around it. This creates a three-dimensional arrangement of carbon atoms throughout the crystal.

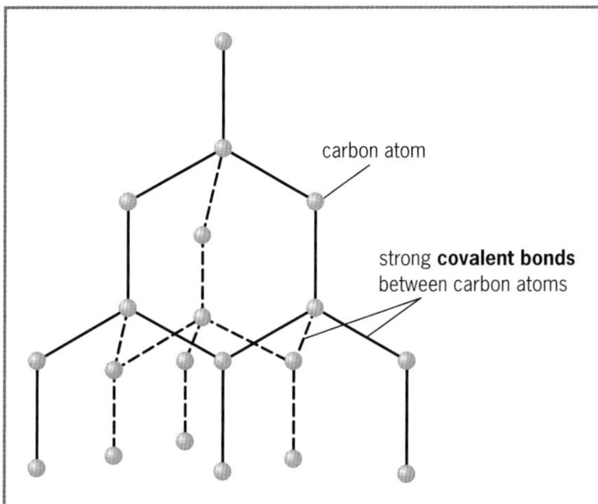

Figure 5.5 *The structure of diamond*

Graphite

In **graphite**, each carbon atom is bonded covalently to **three** others to form **hexagonal rings** of atoms, which are bonded together to form **layers**. The layers have weak forces of attraction between them which hold them together. The fourth electron from each atom becomes **delocalised** and can move within the lattice.

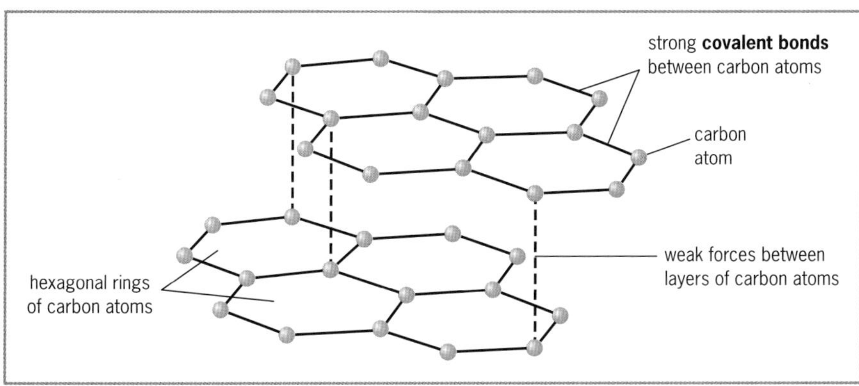

Figure 5.6 *The structure of graphite*

Table 5.6 *A comparison of the properties and uses of sodium chloride, diamond and graphite*

Property	Sodium chloride	Diamond	Graphite
Melting point	**Fairly high** (801 °C) – strong ionic bonds between the ions need a lot of heat energy to break.	**Very high** (3550 °C) – strong covalent bonds between the atoms need large amounts of heat energy to break.	**Very high** (3600 °C) – strong covalent bonds between the atoms need large amounts of heat energy to break. Used to make **crucibles** to hold molten metals.
Solubility in water	**Soluble** – the partial negative ends of polar water molecules attract the positive Na^+ ions and the partial positive ends attract the negative Cl^- ions. This pulls the ions out of the lattice causing the crystal to dissolve.	**Insoluble** – polar water molecules do not attract the carbon atoms out of the lattice.	**Insoluble** – polar water molecules do not attract the carbon atoms out of the lattice.
Conductivity	**Only** conducts electricity when **molten** or **dissolved** in water – the ionic bonds have broken and the ions are free to move. **Does not** conduct electricity in the **solid** state – the ionic bonds hold the ions together so they cannot move.	**Does not** conduct electricity – all the valence electrons are shared between carbon atoms and none are free to move.	**Does** conduct electricity in the **solid** state – the fourth valence electron from each carbon atom is delocalised and free to move. Used to make **electrodes** for use in electrolysis.
Hardness	**Hard** – strong ionic bonds exist between the ions throughout the structure. **Brittle** – if pressure is applied the layers of ions are displaced slightly and ions with the same charges then repel each other and break the lattice apart.	**Extremely hard** – strong covalent bonds exist between the carbon atoms throughout the structure. Used in the tips of **cutting tools** and **drill bits.**	**Soft and flaky** – weak forces exist between the layers of carbon atoms allowing the layers to flake off. Used as the **'lead'** in pencils – the layers slip off and leave dark marks on the paper.
Lubricating power	**None** – the ions are in layers, but ionic bonds hold the layers together so they cannot slide over each other.	**None** – the atoms are bonded covalently throughout the structure and cannot slide over each other.	**Extremely good** – weak forces between the layers of atoms allow the layers to slide over each other. Used as a **lubricant**.

Revision questions

1 Why do elements form compounds?

2 Name the THREE main types of chemical bonding.

3 Determine the formula of EACH of the following compounds and state how EACH would be bonded:

a zinc chloride
b magnesium phosphate
c silicon tetrafluoride
d carbon disulfide
e ammonium carbonate
f aluminium hydroxide
g potassium sulfate

4 Use dot and cross diagrams to show how EACH of the following compounds is formed. Show all electron shells in **a** and **b**, and only the valence electrons in **c** and **d**.

a sodium oxide
b phosphorus trifluoride
c calcium nitride
d methane (CH_4)

(atomic numbers: Na = 11; O = 8; P = 15; F = 9; Ca = 20; N = 7; C = 6; H = 1)

5 Describe the bonding in the metal, magnesium.

6 Explain why a typical metal:

a has a high melting point
b is a good conductor of electricity
c is malleable

7 Using melting point, solubility and conductivity, distinguish between ionic solids and simple molecular solids.

8 Define the term 'allotropy'.

9 By referring to their structure, explain EACH of the following:

a sodium chloride is soluble in water
b diamond is used in cutting tools
c graphite conducts electricity
d graphite is used as the 'lead' in pencils

6 Chemical equations

Chemical equations are used to represent chemical reactions. Reactants and products are shown using symbols and formulae. **Reactants**, separated by plus signs, are shown on the **left. Products**, also separated by plus signs, are shown on the **right**. The reactants and products are separated by an **arrow** and any **conditions** required for the reaction, such as a particular temperature, may be given on the arrow.

$$\text{reactants} \xrightarrow{\text{conditions}} \text{products}$$

Writing balanced equations

When writing an equation it must **balance** so that the **total number** of atoms of **each** element on each side of the arrow is the **same**.

Step 1: Write the equation in **words**:

$$\text{magnesium oxide} + \text{hydrochloric acid} \longrightarrow \text{magnesium chloride} + \text{water}$$

Step 2: Write the **chemical formula** for each reactant and product:

$$\mathbf{MgO + HCl \longrightarrow MgCl_2 + H_2O}$$

Step 3: Show the **physical state** of each reactant and product by placing a **state symbol** after each formula:

solid **(s)**; liquid **(l)**; gas **(g)**; aqueous solution **(aq)**

$$MgO\mathbf{(s)} + HCl\mathbf{(aq)} \longrightarrow MgCl_2\mathbf{(aq)} + H_2O\mathbf{(l)}$$

Step 4: Write down the **total number atoms** of each element in the **reactants** and in the **products**:

$$MgO(s) + HCl(aq) \longrightarrow MgCl_2(aq) + H_2O(l)$$

Reactants	**Products**
Mg = 1	Mg = 1
O = 1	O = 1
H = 1	H = 2
Cl = 1	Cl = 2

Step 5: **Balance** the equation. Do this by placing **simple whole numbers** (or **'coefficients'**) in front of formulae to alter the **proportions** of the reactants and products – **never alter formulae**:

H and **Cl** do not balance. **Balance** by placing a **2** in front of **HCl**.

$$MgO(s) + \mathbf{2}HCl(aq) \longrightarrow MgCl_2(aq) + H_2O(l)$$

Reactants	**Products**
Mg = 1	Mg = 1
O = 1	O = 1
H = ~~1~~ **2**	H = 2
Cl = ~~1~~ **2**	Cl = 2

The equation now **balances**.

Example

Step 1:	sodium	+	water	⟶	sodium hydroxide	+	hydrogen
Step 2:	**Na**	+	$\mathbf{H_2O}$	⟶	**NaOH**	+	$\mathbf{H_2}$
Step 3:	Na**(s)**	+	H_2O**(l)**	⟶	NaOH**(aq)**	+	H_2**(g)**
Step 4:	Na(s)	+	H_2O(l)	⟶	NaOH(aq)	+	H_2(g)

Reactants	**Products**
Na = 1	Na = 1
O = 1	O = 1
H = 2	H = 1 + 2 = 3

Step 5: **H** does not balance. **Balance** by placing a **2** in front of $\mathbf{H_2O}$ and a **2** in front of **NaOH**:

$$Na(s) + 2H_2O(l) \longrightarrow 2NaOH(aq) + H_2(g)$$

Reactants	**Products**
Na = 1	Na = ~~1~~ **2**
O = ~~1~~ **2**	O = ~~1~~ **2**
H = ~~2~~ **4**	H = ~~3~~ 2 + 2 = **4**

Na now does not balance. **Balance** by placing a **2** in front of **Na**:

$$2Na(s) + 2H_2O(l) \longrightarrow 2NaOH(aq) + H_2(g)$$

Reactants	**Products**
Na = ~~1~~ **2**	Na = 2
O = 2	O = 2
H = 4	H = 4

The equation now **balances**.

Useful tips for writing and balancing equations

- For an element in its **free state**, if it is one of the seven common elements that exist as **diatomic molecules**, use its **formula** (H_2, N_2, O_2, F_2, Cl_2, Br_2 and I_2). For any other element in its free state, use its **atomic symbol**.
- Balance the elements in the product **immediately after** the arrow first.
- Balance any **polyatomic** ion which appears unchanged from one side to the other as a unit, e.g. if SO_4^{2-} appears on both sides, consider it as a single unit.
- If hydrogen or oxygen is present in any compound, balance **hydrogen** second from last and balance **oxygen** last.
- Leave elements in their **free state** until the very last to balance, e.g. Mg, Cu, Cl_2, H_2, O_2.
- Check the coefficients to make sure they are in the **lowest** possible ratio.

Example

Balance the following equation:

$$Al(s) + H_2SO_4(aq) \longrightarrow Al_2(SO_4)_3(aq) + H_2(g)$$

Reactants	**Products**
Al = 1	Al = 2
H = 2	H = 2
SO_4 = 1	SO_4 = 3

Start with product immediately after the arrow. **Al** does not balance. **Balance** by placing a **2** in front of the **Al**. $\mathbf{SO_4}$ does not balance. **Balance** by placing a **3** in front of the $\mathbf{H_2SO_4}$:

$$2Al(s) + 3H_2SO_4(aq) \longrightarrow Al_2(SO_4)_3(aq) + H_2(g)$$

Reactants	**Products**
Al = ~~1~~ **2**	Al = 2
H = ~~2~~ **6**	H = 2
SO_4 = ~~1~~ **3**	SO_4 = 3

H now does not balance. **Balance** by placing a **3** in front of the $\mathbf{H_2}$:

$$2Al(s) + 3H_2SO_4(aq) \longrightarrow Al_2(SO_4)_3(aq) + 3H_2(g)$$

Reactants	**Products**
Al = 2	Al = 2
H = 6	H = ~~2~~ **6**
SO_4 = 3	SO_4 = 3

The equation is now **balanced**.

To determine state symbols of ionic compounds

In equations, if the compound is **soluble**, it would usually be given the state symbol **(aq)**. If it is **insoluble** it would always be given the state symbol **(s)**. Most **ionic compounds** *are* **soluble** in water; however, some are insoluble. It is extremely important to learn the rules to determine the solubility of ionic compounds. These rules are given in Tables 6.1 and 6.2.

Table 6.1 *Common ionic compounds which are soluble in water*

Compounds	Solubility in water	Exceptions
Potassium, sodium and **ammonium** compounds	Soluble	None.
Nitrates	Soluble	None.
Chlorides	Soluble	**Silver chloride (AgCl)** is insoluble. **Lead(II) chloride** ($\mathbf{PbCl_2}$) is insoluble in cold water but moderately soluble in hot water.
Sulfates	Soluble	**Lead(II) sulfate** ($\mathbf{PbSO_4}$) and **barium sulfate** ($\mathbf{BaSO_4}$) are insoluble. **Calcium sulfate** ($\mathbf{CaSO_4}$) is slightly soluble.
Ethanoates	Soluble	None.
Hydrogencarbonates	Soluble	None.

Table 6.2 *Common ionic compounds which are insoluble in water*

Compounds	Solubility in water	Exceptions
Carbonates	Insoluble	**Potassium carbonate** (K_2CO_3), **sodium carbonate** (Na_2CO_3) and **ammonium carbonate** ($(NH_4)_2CO_3$) are soluble.
Phosphates	Insoluble	**Potassium phosphate** (K_3PO_4), **sodium phosphate** (Na_3PO_4) and **ammonium phosphate** ($(NH_4)_3PO_4$) are soluble.
Hydroxides	Insoluble	**Potassium hydroxide** (KOH), **sodium hydroxide** (NaOH) and **ammonium hydroxide** (NH_4OH) are soluble. **Calcium hydroxide** ($Ca(OH)_2$) is slightly soluble.
Metal oxides	Insoluble	**Potassium oxide** (K_2O), **sodium oxide** (Na_2O) and **calcium oxide** (CaO) react with water to form the equivalent soluble hydroxides.

Ionic equations

Ionic equations show only the atoms or ions which **change** during a reaction. These are the atoms or ions which actually **take part** in the reaction:

- Two ions in solution may join to form a **precipitate** (an insoluble compound within the solution).
- Two ions may form a **covalent** compound (a compound composed of molecules).
- An ion may be **discharged** (converted to an atom).
- An atom of an element in its free state may be **ionised** (converted to an ion).

Ions which remain **unchanged** during a reaction are known as **spectator ions** and are not included in an ionic equation. Spectator ions remain **in solution** during the reaction.

Writing ionic equations

The four steps given below should be followed when writing ionic equations.

Step 1: Write the **balanced equation** for the reaction:

$$AgNO_3(aq) + KCl(aq) \longrightarrow AgCl(s) + KNO_3(aq)$$

Step 2: Rewrite the equation showing any **ions** in **solution** as individual ions:

$$Ag^+(aq) + NO_3^-(aq) + K^+(aq) + Cl^-(aq) \longrightarrow AgCl(s) + K^+(aq) + NO_3^-(aq)$$

Step 3: Delete the ions which remain **unchanged**, i.e. the **NO_3^-(aq)** and **K^+(aq)** ions remain in solution:

$$Ag^+(aq) + \cancel{NO_3^-(aq)} + \cancel{K^+(aq)} + Cl^-(aq) \longrightarrow AgCl(s) + \cancel{K^+(aq)} + \cancel{NO_3^-(aq)}$$

Step 4: Rewrite the **ionic equation** showing the ions which **change**, i.e. the **Ag^+(aq)** and **Cl^-(aq)** ions form an insoluble precipitate of silver chloride, **AgCl(s)**:

$$Ag^+(aq) + Cl^-(aq) \longrightarrow AgCl(s)$$

Example

Step 1: $Na_2CO_3(aq) + H_2SO_4(aq) \longrightarrow Na_2SO_4(aq) + CO_2(g) + H_2O(l)$

Step 2:

$$2Na^+(aq) + CO_3^{2-}(aq) + 2H^+(aq) + SO_4^{2-}(aq) \longrightarrow 2Na^+(aq) + SO_4^{2-}(aq) + CO_2(g) + H_2O(l)$$

Step 3: Delete the **$Na^+(aq)$** and **$SO_4^{2-}(aq)$** ions:

$$\cancel{2Na^+(aq)} + CO_3^{2-}(aq) + 2H^+(aq) + \cancel{SO_4^{2-}(aq)} \longrightarrow \cancel{2Na^+(aq)} + \cancel{SO_4^{2-}(aq)} + CO_2(g) + H_2O(l)$$

Step 4: Ionic equation:

$$CO_3^{2-}(aq) + 2H^+(aq) \longrightarrow CO_2(g) + H_2O(l)$$

Point to note:

If the coefficients are not in the lowest ratio, cancel to the **lowest possible ratio**, for example:

The reaction between sodium hydroxide solution and sulfuric acid:

$$2NaOH(aq) + H_2SO_4(aq) \longrightarrow Na_2SO_4(aq) + 2H_2O(l)$$

Following **steps 2 to 4**, the ionic equation for the reaction works out as follows:

$$2OH^-(aq) + 2H^+(aq) \longrightarrow 2H_2O(l)$$

This cancels to:

$$OH^-(aq) + H^+(aq) \longrightarrow H_2O(l)$$

Revision questions

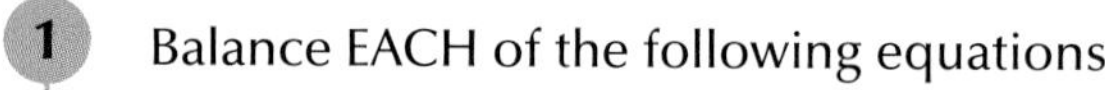

1 Balance EACH of the following equations:

a $Br_2(aq) + KI(aq) \longrightarrow KBr(aq) + I_2(aq)$

b $Fe(s) + Cl_2(g) \longrightarrow FeCl_3(s)$

c $Al(s) + H_2SO_4(aq) \longrightarrow Al_2(SO_4)_3(aq) + H_2(g)$

d $C_2H_4(g) + O_2(g) \longrightarrow CO_2(g) + H_2O(g)$

e $NaOH(aq) + (NH_4)_2SO_4(aq) \longrightarrow Na_2SO_4(aq) + NH_3(g) + H_2O(l)$

2 State whether EACH of the following ionic compounds is soluble or insoluble in water:

a silver nitrate

b potassium phosphate

c zinc hydroxide

d aluminium sulfate

e lead(II) chloride

f copper(II) oxide

g calcium carbonate

h sodium ethanoate

3 Write a balanced chemical equation for EACH of the following reactions:

a The reaction between magnesium hydroxide and nitric acid ($HNO_3(aq)$) to produce magnesium nitrate and water.

b The reaction between lead(II) nitrate solution and sodium chloride solution to form lead(II) chloride and sodium nitrate.

c The reaction between calcium hydrogencarbonate and hydrochloric acid ($HCl(aq)$) to form calcium chloride, carbon dioxide and water.

d The formation of zinc oxide, nitrogen dioxide (NO_2) gas and oxygen when solid zinc nitrate is heated.

4 Write the ionic equation for the reactions in **1a**, **1c**, **1e**, **3b** and **3c** above.

7 Types of chemical reaction

There are **seven** main types of chemical reaction:

Synthesis reactions

During a **synthesis reaction**, two or more substances combine chemically to produce a **single product**:

A + B → AB

e.g. $2Al(s) + 3Cl_2(g) \longrightarrow 2AlCl_3(s)$

Decomposition reactions

During a **decomposition reaction**, a single reactant is **broken down** into two or more products. This can be carried out by heating the compound (**thermal decomposition**), or by passing an electric current through the compound in the liquid state or dissolved in aqueous solution (**electrolysis**).

AB → A + B

e.g. $MgCO_3(s) \xrightarrow{\text{heat}} MgO(s) + CO_2(g)$

Single displacement reactions

During a **single displacement reaction**, an element in its **free state** takes the place of (displaces) another element in the compound. A **more reactive** element always displaces a less reactive element. There are two types:

- A **metal** may displace the hydrogen from an acid or another metal from a compound.

  A + BY → AY + B

  e.g. $Mg(s) + 2HCl(aq) \longrightarrow MgCl_2(aq) + H_2(g)$

  $Zn(s) + CuSO_4(aq) \longrightarrow ZnSO_4(aq) + Cu(s)$

  Magnesium is above hydrogen in the reactivity series (see p. 158) and zinc is more reactive than copper.

- A **non-metal** may displace another non-metal from a compound.

  X + AY → AX + Y

  e.g. $Cl_2(g) + 2KI(aq) \longrightarrow 2KCl(aq) + I_2(aq)$

  Chlorine is more reactive than iodine.

Ionic precipitation reactions

During an **ionic precipitation** reaction, two compounds in solution **exchange ions** to form an **insoluble precipitate** and another soluble compound. These reactions are sometimes called **double displacement reactions**.

AX + BY → AY + BX

e.g. $Pb(NO_3)_2(aq) + 2NaCl(aq) \longrightarrow PbCl_2(s) + 2NaNO_3(aq)$

Oxidation-reduction reactions

During an **oxidation-reduction reaction**, one reactant is **oxidised** and the other is **reduced**. These reactions are also called **redox reactions** (see p. 82).

e.g. $CH_4(g) + 2O_2(g) \longrightarrow CO_2(g) + 2H_2O(g)$

Neutralisation reactions

During **neutralisation reactions**, a **base** reacts with an **acid** to produce a salt and water (see p. 77).

e.g. $2KOH(aq) + H_2SO_4(aq) \longrightarrow K_2SO_4(aq) + 2H_2O(l)$

Reversible reactions

During a **reversible reaction**, the direction of the chemical change can be easily **reversed**. The products can react to produce the original reactants again.

$$A + B \rightleftharpoons C + D$$

e.g. $NH_4Cl(s) \rightleftharpoons NH_3(g) + HCl(g)$

If ammonium chloride is **heated**, it decomposes into ammonia gas and hydrogen chloride gas:

$$NH_4Cl(s) \longrightarrow NH_3(g) + HCl(g)$$

If the ammonia gas and hydrogen chloride gas are **cooled**, or mixed at room temperature, they react to form ammonium chloride:

$$NH_3(g) + HCl(g) \longrightarrow NH_4Cl(s)$$

The reaction is, therefore, **reversible.**

In many reversible reactions, the reaction contains a **mixture** of reactants and products because it proceeds in **both directions** at the same time. Most reactions can only proceed in one direction, so they are not reversible.

Revision questions

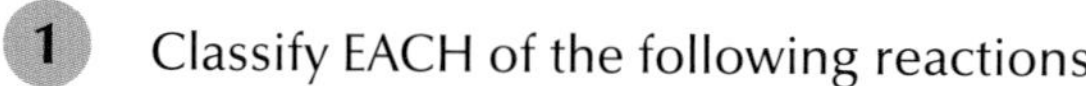

1 Classify EACH of the following reactions:

a The reaction between calcium and hydrochloric acid to form calcium chloride and hydrogen.

b Iron reacting with oxygen to produce iron(III) oxide.

c The reaction between sodium hydroxide and nitric acid to produce sodium nitrate and water.

d The formation of calcium oxide and steam when calcium hydroxide is heated.

e The reaction between barium nitrate solution and potassium sulfate solution to make barium sulfate and potassium nitrate.

f The formation of sodium chloride and bromine when chlorine gas is bubbled into sodium bromide solution.

8 The mole concept

Atoms, molecules and formula units are extremely small and cannot be counted directly. The **mole** is the **unit of amount** used in Chemistry to provide a bridge between atoms, molecules or formula units and the amount of chemical substances that can be worked with in the laboratory. The **mole** represents a **specific number** of atomic sized particles, just as a pair, dozen or gross represent specific numbers of objects.

Relative atomic, molecular and formula masses

Because the mass of atoms is extremely small, **relative atomic mass** is used to **compare** their masses.

A carbon-12 atom has been assigned a mass of **12.00 atomic mass units** or **amu**. This means that 1/12th the mass of a carbon-12 atom has a mass of **1.00 amu**. Relative atomic mass **compares** the masses of atoms to this value:

Relative atomic mass (A_r) *is the average mass of* ***one atom of an element*** *compared to one-twelfth the mass of an atom of carbon-12.*

The word 'average' is used in the definition because relative atomic mass takes into account the relative abundance of each **isotope**. Relative atomic mass has **no units** because it is a comparative value.

The masses of covalent and ionic compounds are also compared to the carbon-12 atom:

- The term **relative molecular mass** is used when referring to **covalently bonded** elements or compounds:

  Relative molecular mass (M_r) *is the average mass of* ***one molecule of an element or compound*** *compared to one-twelfth the mass of an atom of carbon-12.*

- The term **relative formula mass** is used when referring to **ionic compounds**:

  Relative formula mass *is the average mass of* ***one formula unit of an ionic compound*** *compared to one-twelfth the mass of an atom of carbon-12.*

Calculating relative atomic, molecular and formula masses

The **relative atomic mass** of each element, to the nearest whole number, can be found in the **periodic table** on p. 196.

Relative molecular mass and **relative formula mass** are calculated by **adding** together the relative atomic masses of all the elements present in the compound.

Examples

Relative atomic masses:

- Aluminium, **Al** = **27**
- Sulfur, **S** = **32**
- Iron, **Fe** = **56**

Relative molecular masses:

- Hydrogen, $\mathbf{H_2}$

  H_2 consists of two H atoms.

  ∴ relative molecular mass of $H_2 = (2 \times 1) = \mathbf{2}$
- Sulfur trioxide, $\mathbf{SO_3}$

  SO_3 consists of one S atom and three O atoms.

  ∴ relative molecular mass of $SO_3 = 32 + (3 \times 16) = \mathbf{80}$

- Glucose, $C_6H_{12}O_6$

  Relative molecular mass of $C_6H_{12}O_6 = (6 \times 12) + (12 \times 1) + (6 \times 16) = \mathbf{180}$

Relative formula masses:

- Potassium sulfide, K_2S

  Relative formula mass of $K_2S = (2 \times 39) + 32 = \mathbf{110}$

- Magnesium nitrate, $Mg(NO_3)_2$

  Relative formula mass of $Mg(NO_3)_2 = 24 + (2 \times 14) + (2 \times 3 \times 16) = \mathbf{148}$

The mole

A **mole** is the amount of a substance that contains the **same number** of **particles** as there are atoms in 12.00 g of carbon-12. It was found that 12.00 g of carbon-12 contains **6.0×10^{23} atoms** of carbon-12, therefore, **a mole** represents **6.0×10^{23}**.

*A **mole** is the amount of a substance that contains **6.0×10^{23} particles** of the substance.*

In the above definition:

- 'amount' can be the **mass** of a substance, or the **volume** of a substance if it is a **gas**.
- 'particles' can be **atoms, molecules, formula units** or **ions**.

The number **6.0×10^{23}** is known as **Avogadro's constant**, or N_A.

The mole and mass

Just as it was found that 12.00 g of carbon-12 contains 6.0×10^{23} carbon-12 atoms, it was also found that:

- **27 g** of aluminium (Al) contains 6.0×10^{23} Al atoms.
- **80 g** of sulfur trioxide (SO_3) contains 6.0×10^{23} SO_3 molecules.
- **148 g** of magnesium nitrate ($Mg(NO_3)_2$) contains 6.0×10^{23} $Mg(NO_3)_2$ formula units.

Looking at the **masses** given above:

- Each mass has the same numerical value as the relative atomic, molecular or formula mass of the element or compound.
- Each is the mass of **6.0×10^{23}** particles, or **one mole** of particles of the substance.

It therefore follows that **one mole** of a substance has a mass equal to the relative atomic, molecular or formula mass expressed in **grams**.

***Molar mass (M)** is the mass, in grams, of one mole of a chemical substance.*

The molar mass of an element or compound is given the unit **grams per mole** or **g mol^{-1}**. For example, the molar mass of carbon is **12 g mol^{-1}**.

Therefore, the **molar mass** of an element or compound is the relative atomic, molecular or formula mass amount expressed in **grams per mole**.

Examples

Molar mass of $\mathbf{CO_2}$ = $12 + (2 \times 16)$ g mol^{-1} = **44 g mol^{-1}**

Molar mass of $\mathbf{CaSO_4}$ = $40 + 32 + (4 \times 16)$ g mol^{-1} = **136 g mol^{-1}**

Molar mass of $\mathbf{Al_2(CO_3)_3}$ = $(2 \times 27) + (3 \times 12) + (3 \times 3 \times 16)$ g mol^{-1} = **234 g mol^{-1}**

Molar mass (mass of one mole) can be used to convert a given mass of an element or compound to number of moles, or to convert a given number of moles of an element or compound to mass.

Two conversions

a Given mass of element or compound to number of moles:

$$\text{number of moles} = \frac{\textbf{given mass}}{\text{mass of one mole (molar mass)}}$$

b Given number of moles to mass:

$$\text{mass} = \textbf{given number of moles} \times \text{mass of one mole (molar mass)}$$

Sample questions

1 How many moles are there in 27 g of water?

Mass of 1 mol H_2O = (2 × 1) + 16 g = **18 g**

$\therefore$ number of moles in **27 g H_2O** $= \frac{27}{18}$ mol (using equation **a** above)

$= \underline{\textbf{1.5 mol}}$

2 What is the mass of 0.25 mol of iron(III) sulfate?

Mass of 1 mol $Fe_2(SO_4)_3$ = (2 × 56) + (3 × 32) + (3 × 4 × 16) g = **400 g**

$\therefore$ mass of **0.25 mol $Fe_2(SO_4)_3$** = 0.25 × 400 g (using equation **b** above)

$= \underline{\textbf{100 g}}$

The mole and number of particles

The fact that the number of particles in one mole is always **6.0×10^{23}** can be used to convert a given number of particles in a substance to number of moles, or to convert a given number of moles to the number of particles.

Two conversions

c Given number of particles to number of moles:

$$\text{number of moles} = \frac{\textbf{given number of particles}}{6.0 \times 10^{23}}$$

d Given number of moles to number of particles:

$$\text{number of particles} = \textbf{given number of moles} \times 6.0 \times 10^{23}$$

The **type** of particle in a substance depends on the nature of the substance:

- If the substance is an **element**, e.g. a metal or a noble gas, the particles are individual **atoms**.
- If the substance is a **molecular element**, e.g. nitrogen, N_2, or a **covalent compound**, e.g. carbon dioxide, CO_2, the particles are **molecules** made up of **atoms**.
- If the substance is an **ionic compound**, e.g. potassium chloride, KCl, the particles are **formula units** made up of **ions**.

Sample questions

1 How many moles are in 1.8×10^{23} molecules of nitrogen?

1 mol N_2 contains 6.0×10^{23} N_2 molecules

∴ number of moles in **1.8×10^{23} N_2 molecules** $= \frac{1.8 \times 10^{23}}{6.0 \times 10^{23}}$ mol (using equation **c** above)

$= \underline{\mathbf{0.3\ mol}}$

2 Calculate the number of magnesium oxide formula units in 0.75 mol of magnesium oxide.

1 mol MgO contains 6.0×10^{23} MgO formula units

∴ **0.75 mol MgO** contains $0.75 \times 6.0 \times 10^{23}$ MgO formula units (using equation **d** above)

$= \underline{\mathbf{4.5 \times 10^{23}\ MgO\ formula\ units}}$

3 How many chloride ions are there in 0.2 mol of aluminium chloride?

1 mol $AlCl_3$ contains 6.0×10^{23} $AlCl_3$ formula units

∴ **0.2 mol $AlCl_3$** contains $0.2 \times 6.0 \times 10^{23}$ $AlCl_3$ formula units (using equation **d**)

$= \mathbf{1.2 \times 10^{23}\ AlCl_3\ formula\ units}$

1 $AlCl_3$ formula unit contains 3 Cl^- ions

∴ **1.2×10^{23} $AlCl_3$ formula units** contain $3 \times 1.2 \times 10^{23}$ Cl^- ions

$= \underline{\mathbf{3.6 \times 10^{23}\ Cl^-\ ions}}$

The mole, mass and number of particles

The calculation of moles and mass, and moles and number of particles can be combined.

Combined conversions

mass of element or compound $\underset{b}{\overset{a}{\rightleftharpoons}}$ **number of moles** $\underset{d}{\overset{c}{\leftleftharpoons}}$ number of particles

Sample questions

1 How many ammonia molecules are there in 4.25 g of ammonia?

Mass of 1 mol NH_3 = 14 + (3 × 1) g = **17g**

∴ number of moles in **4.25 g NH_3** $= \frac{4.25}{17}$ mol (using equation **a**)

$= \mathbf{0.25\ mol}$

1 mol NH_3 contains 6.0×10^{23} NH_3 molecules

∴ **0.25 mol NH_3** contains $0.25 \times 6.0 \times 10^{23}$ NH_3 molecules (using equation **d**)

$= \underline{\mathbf{1.5 \times 10^{23}\ NH_3\ molecules}}$

2 Calculate the mass of 2.4×10^{22} calcium nitrate formula units.

1 mol $Ca(NO_3)_2$ contains 6.0×10^{23} $Ca(NO_3)_2$ formula units

∴ number of moles in **2.4×10^{22} $Ca(NO_3)_2$ formula units** $= \frac{2.4 \times 10^{22}}{6.0 \times 10^{23}}$ mol (using equation **c**)

$= \mathbf{0.04\ mol}$

Mass of 1 mol $Ca(NO_3)_2$ = 40 + (2 × 14) + (2 × 3 × 16) g = **164 g**

∴ mass of **0.04 mol $Ca(NO_3)_2$** = 0.04 × 164 g (using equation **b**)

$= \underline{\mathbf{6.56\ g}}$

The mole and volumes of gases

Avogadro's Law states that equal volumes of all gases, under the same conditions of temperature and pressure, contain the same number of molecules.

If the number of molecules in each gas is **6.0 × 10²³** (**1 mol**) it follows that one mole of all gases, at the same temperature and pressure, occupy the **same volume**.

Molar volume (V_m) *is the volume occupied by one mole (6.0 × 10²³ molecules) of a gas.*

Molar volume depends on temperature and pressure:

- At **standard temperature and pressure** (**stp**), where temperature is 0 °C (273 K) and pressure is 101.3 kPa (1 atmosphere), the volume of one mole of a gas is **22.4 dm³** or **22 400 cm³**.
- At **room temperature and pressure** (**rtp**), where temperature is 25 °C (298 K) and pressure is 101.3 kPa (1 atmosphere), the volume of one mole of a gas is **24.0 dm³** or **24 000 cm³**.

Molar volume can be used to convert a given volume of a gas to number of moles, or to convert a given number of moles of a gas to volume.

Two conversions

e Given volume of gas to number of moles:

$$\text{number of moles} = \frac{\textbf{given volume of gas}}{\text{volume of one mole at stp or rtp (molar volume)}}$$

f Given number of moles to volume of gas:

volume of gas = **given number of moles** × volume of one mole at stp or rtp (molar volume)

Sample questions

1 How many moles are there in 3.36 dm³ of carbon dioxide at stp?

Volume of 1 mol CO_2 at stp = 22.4 dm³

∴ number of moles in **3.36 dm³ CO_2** $= \frac{3.36}{22.4}$ mol (using equation **e** above)

= **0.15 mol**

2 What volume is occupied by 0.2 mol of chlorine gas at rtp?

Volume of 1 mol Cl_2 at rtp = 24.0 dm³

∴ volume of **0.2 mol Cl_2** = 0.2 × 24.0 dm³ (using equation **f** above)

= **4.8 dm³**

The mole, mass, number of particles and gas volume

The calculations of moles and mass, moles and number of particles, and moles and gas volumes can be combined.

Combined conversions

mass of element or compound ⇌ (a → / ← b) **number of moles** ⇌ (← c / d →) number of particles

number of moles ⇅ (e ↑ / f ↓) volume of gas

Sample questions

1 What is the volume of 12.8 g of oxygen at stp?

Mass of 1 mol O_2 = (2×16) g = **32 g**

$\therefore$ number of moles in **12.8 g O_2** $= \frac{12.8}{32}$ mol (using equation **a**)

$= \mathbf{0.4\ mol}$

Volume of 1 mol O_2 at stp = 22.4 dm^3

$\therefore$ volume of **0.4 mol O_2** = $0.4 \times 22.4\ dm^3$ (using equation **f**)

$= \underline{\mathbf{8.96\ dm^3}}$

2 How many sulfur trioxide molecules are there in 720 cm^3 of sulfur trioxide gas at rtp?

Volume of 1 mol SO_3 at rtp = 24 000 cm^3

$\therefore$ number of moles in **720 cm^3 SO_3** $= \frac{720}{24\ 000}$ mol (using equation **e**)

$= \mathbf{0.03\ mol}$

1 mol SO_3 contains 6.0×10^{23} SO_3 molecules

$\therefore$ **0.03 mol SO_3** contains $0.03 \times 6.0 \times 10^{23}$ SO_3 molecules (using equation **d**)

$= \underline{\mathbf{1.8 \times 10^{22}\ SO_3\ molecules}}$

Revision questions

1 Define EACH of the following:

a relative atomic mass **b** a mole **c** molar mass **d** molar volume

2 **a** What is the mass of 0.3 mol of ammonium phosphate?

b How many moles are there in 3.2 g of copper(II) sulfate?

c How many moles of aluminium oxide would contain 2.4×10^{22} aluminium oxide formula units?

d How many carbon dioxide molecules are there in 11 g of carbon dioxide?

3 State Avogadro's Law.

4 **a** How many moles are there in 560 cm^3 of oxygen at stp?

b What volume would be occupied by 0.15 mol of carbon monoxide at rtp?

c What is the volume of 3.4 g of ammonia at rtp?

d You have a container of hydrogen with a volume of 1.68 dm^3. How many hydrogen molecules would it contain at stp?

The mole and chemical formulae

When considered in terms of moles, a **chemical formula** shows how many **moles of each element** are combined to form one mole of a compound. For example, CO_2 represents 1 mol of carbon atoms combined with 2 mol of oxygen atoms.

Chemical formulae can be written in two main ways:

- The **empirical formula**. This gives the simplest **whole number mole ratio** between the atoms or ions present in the compound. **Ionic compounds** are always represented by **empirical formulae**.
- The **molecular formula**. This gives the **actual number of moles** of atoms of each element present in one mole of the compound. **Covalent compounds** are represented by **molecular formulae**.

The empirical and molecular formulae of **covalent compounds** may be the same, e.g. carbon dioxide, CO_2. However, in some compounds, the molecular formula is a simple **whole number multiple** of the empirical formula, e.g. the molecular formula of glucose is $C_6H_{12}O_6$ and its empirical formula is CH_2O.

If the **proportions** of the elements, by mass, in a compound are known, then its empirical formula can be determined. If the molecular formula is different from the empirical formula and the relative molecular mass or molar mass is known, the molecular formula can be determined.

Sample questions

1 When analysed, a compound of mass 50.0 g was found to contain 18.3 g of sodium, 12.7 g of sulfur and an unknown mass of oxygen. Determine the empirical formula of the compound.

Mass of **oxygen** in the compound = 50.0 – (18.3 + 12.7) g = **19.0 g**

Element	Na	S	O
Mass	18.3 g	12.7 g	19.0 g
Mass of 1 mole	23 g	32 g	16 g
Number of moles	$\frac{18.3}{23}$ mol = **0.796 mol**	$\frac{12.7}{32}$ mol = **0.397 mol**	$\frac{19.0}{16}$ mol = **1.188 mol**
Simplest mole ratio (dividing by the smallest number of moles)	**2 mol**	**1 mol**	**3 mol**

Empirical formula is $\mathbf{Na_2SO_3}$

2 On analysis, a compound was found to contain 54.5% carbon, 9.1% hydrogen and 36.4% oxygen. Determine the molecular formula of this compound if its relative molecular mass was 88.

Element	C	H	O
Percentage	54.5%	9.1%	36.4%
Mass in 100 g	54.5 g	9.1 g	36.4 g
Mass of 1 mole	12 g	1 g	16 g
Number of moles	$\frac{54.5}{12}$ mol = **4.542 mol**	$\frac{9.1}{1}$ mol = **9.1 mol**	$\frac{36.4}{16}$ mol = **2.275 mol**
Simplest mole ratio	**2 mol**	**4 mol**	**1 mol**

Empirical formula is $\mathbf{C_2H_4O}$

To determine the **molecular formula:**

Relative molecular mass of C_2H_4O = (2 × 12) + (4 × 1) + 16 = **44**

Relative molecular mass of compound = **88**

and ratio between relative molecular masses = $\frac{88}{44}$ = **2**

∴ the molecular formula is **2** × the empirical formula

Molecular formula of the compound is $\mathbf{C_4H_8O_2}$

Percentage composition

Percentage composition is the percentage, by mass, of each element in a compound. This can be calculated once the formula of a compound is known.

Sample question

Calculate the percentage, by mass, of carbon in aluminium carbonate, $Al_2(CO_3)_3$.

Mass of 1 mol $Al_2(CO_3)_3$ = (2 × 27) + (3 × 12) + (3 × 3 × 16) g = **234 g**

Mass of **carbon** in 1 mol $Al_2(CO_3)_3$ = 3 × 12 g = **36 g**

∴ percentage of carbon in $Al_2(CO_3)_3 = \frac{36}{234} \times 100\%$

= **<u>15.38%</u>**

The mole and solutions

The **concentration of a solution** is a measure of how much solute is dissolved in a fixed volume of the **solution**. The volume usually used is **1 dm³** (1000 cm³). Concentration can be expressed in two ways:

- **Mass concentration** gives the **mass** of solute dissolved in 1 dm³ of solution. The unit is **grams of solute per cubic decimetre** of solution or **g dm⁻³**.
- **Molar concentration** gives the **number of moles** of solute dissolved in 1 dm³ of solution. The unit is **moles of solute per cubic decimetre** of solution (**mol dm⁻³**).

*A **standard solution** is a solution whose concentration is known accurately.*

A standard solution is made in a **volumetric flask**. The flask has an accurate volume which is not always 1 dm³. The solute is weighed out, added to the flask and water is added to the mark on the neck of the flask. The concentration of the solution can then be calculated as in the example below:

Example

A sodium carbonate solution contains 21.2 g of sodium carbonate dissolved in 1 dm³ of solution.

∴ **mass concentration** of the solution = **21.2 g dm⁻³**

And mass of 1 mole Na_2CO_3 = (2 × 23) + 12 + (3 × 16) g = **106 g**

∴ number of moles in 21.2 g $Na_2CO_3 = \frac{21.2}{106}$ mol (using equation **a**)

= **0.2 mol**

∴ **molar concentration** of the solution = **<u>0.2 mol dm⁻³</u>**

Sample questions

1 250 cm³ of potassium hydroxide solution contains 8.4 g of potassium hydroxide. What are the mass and molar concentrations of this solution?

250 cm³ KOH(aq) contains 8.4 g KOH

∴ 1 cm³ KOH(aq) contains $\frac{8.4}{250}$ g KOH

and **1000 cm³ KOH(aq)** contains $\frac{8.4}{250} \times 1000$ g KOH

= **33.6 g KOH**

Mass concentration of the solution = **<u>33.6 g dm⁻³</u>**

Mass of 1 mol KOH = 39 + 16 + 1 g = **56 g**

∴ number of moles in **33.6 g KOH** = $\frac{33.6}{56}$ mol (using equation **a**)

= **0.6 mol**

Molar concentration of the solution = **<u>0.6 mol dm⁻³</u>**

2 How many moles of calcium nitrate are present in 200 cm^3 of a solution which has a molar concentration of 0.4 mol dm^{-3}?

1 dm^3 $Ca(NO_3)_2(aq)$ contains 0.4 mol $Ca(NO_3)_2$

i.e. 1000 cm^3 $Ca(NO_3)_2(aq)$ contains 0.4 mol $Ca(NO_3)_2$

$\therefore$ 1 cm^3 $Ca(NO_3)_2(aq)$ contains $\frac{0.4}{1000}$ mol $Ca(NO_3)_2$

and **200 cm^3 $Ca(NO_3)_2(aq)$** contains $\frac{0.4}{1000} \times 200$ mol $Ca(NO_3)_2$

$= \mathbf{0.08\ mol\ Ca(NO_3)_2}$

3 What mass of sodium hydroxide would be needed to produce 750 cm^3 of sodium hydroxide solution with a concentration of 0.5 mol dm^{-3}?

1000 cm^3 NaOH(aq) contains 0.5 mol NaOH

$\therefore$ **750 cm^3 NaOH(aq)** contains $\frac{0.5}{1000} \times 750$ mol NaOH

$= \mathbf{0.375\ mol\ NaOH}$

Mass of 1 mol NaOH = 23 + 16 + 1 g = **40 g**

$\therefore$ mass of **0.375 mol NaOH** = 0.375 × 40 g (using equation **b**)

= **15.0 g**

Mass of sodium hydroxide required = **15.0 g**

Revision questions

5 **a** On analysis, 56.0 g of a compound was found to contain 23.0 g of potassium, 18.9 g of sulfur and an unknown mass of oxygen. Determine the empirical formula of the compound.

b A compound with a relative molecular mass of 216 was found to contain 22.2% carbon, 3.7% hydrogen and 74.1% bromine. Determine the molecular formula of this compound.

6 What is the percentage, by mass, of oxygen in lead(II) nitrate?

7 What is a standard solution?

8 **a** What mass of potassium carbonate would you need to prepare 250 cm^3 of a solution of concentration 0.2 mol dm^{-3}?

b How many moles of sodium hydroxide are present in 200 cm^3 of a solution which has a concentration of 16.5 g dm^{-3}?

c Determine the mass concentration and molar concentration of a solution of calcium nitrate if 400 cm^3 of the solution contains 32.8 g of calcium nitrate.

The mole and chemical reactions

The Law of Conservation of Matter *states that matter can neither be created nor destroyed during a chemical reaction.*

In any chemical reaction, the total mass of the products is the **same** as the total mass of the reactants. When considered in terms of moles, the **coefficients** in a balanced chemical equation show the number of **moles** of each reactant and product.

Example

$$2Al(s) + 3Cl_2(g) \longrightarrow 2AlCl_3(s)$$

This means:

$$\textbf{2 mol} \text{ of } Al + \textbf{3 mol} \text{ of } Cl_2 \longrightarrow \textbf{2 mol} \text{ of } AlCl_3$$

Converting moles to mass, it means:

$$\textbf{2}(27) \text{ g of } Al + \textbf{3}(2 \times 35.5) \text{ g of } Cl_2 \longrightarrow \textbf{2}(27 + (3 \times 35.5)) \text{ g of } AlCl_3$$

$$\textbf{54 g} \text{ of } Al + \textbf{213 g} \text{ of } Cl_2 \longrightarrow \textbf{267 g} \text{ of } AlCl_3$$

The total mass of the original reactants, aluminium and chlorine, is **267 g** which is the same as the mass of the product, aluminium chloride.

Applying the Law of Conservation of Matter, if the quantity of one reactant or product is known, the quantities of any of the other reactants and products can be calculated.

Steps to follow when answering questions

Step 1: Write the **balanced chemical equation** for the reaction if it has not been given. If the question refers to ions, then write the balanced **ionic equation**.

Step 2: Write the **quantity** of the reactant or product which has been given underneath its formula in the equation and place a **question mark** under the reactant or product whose quantity is being calculated.

Step 3: Convert the **given quantity** of reactant or product to **moles**. This is the **known** reactant or product.

Step 4: Use the balanced equation to determine the **mole ratio** between the known and the unknown reactant or product. The **unknown** reactant or product is the one whose quantity is being calculated.

Step 5: Use the **number of moles** of the known reactant or product found in **step 3** and the **mole ratio** found in **step 4** to calculate the **number of moles** of the unknown.

Step 6: Use the **number of moles** of the unknown reactant or product found in **step 5** and its molar mass or volume to determine its **quantity**.

Example

To determine the maximum mass of sodium sulfate that can be produced when 24.0 g of sodium hydroxide reacts with excess sulfuric acid.

Steps 1 and **2:**

$$2NaOH(aq) + H_2SO_4(aq) \longrightarrow Na_2SO_4(aq) + 2H_2O(l)$$

24.0 g		? mass	

(**24.0 g** under $2NaOH(aq)$; **? mass** under $Na_2SO_4(aq)$)

Mass of **NaOH** is known, mass of $\mathbf{Na_2SO_4}$ is unknown.

Step 3: Find the **number of moles** of the known reactant, i.e. NaOH, using its molar mass and given mass:

Mass of 1 mol NaOH = 23 + 16 + 1 g

= **40 g**

∴ number of moles in **24.0 g NaOH** = $\frac{24.0}{40}$ mol

= **0.6 mol**

Step 4: Use the balanced equation to determine the **mole ratio** between the NaOH and the unknown product, i.e. Na_2SO_4:

2 mol NaOH form **1 mol Na_2SO_4**

Step 5: Use the number of moles of NaOH from **step 3** and the mole ratio from **step 4** to calculate the **number of moles** of Na_2SO_4 produced:

0.6 mol NaOH forms $\frac{1}{2}$ × 0.6 mol Na_2SO_4

= **0.3 mol Na_2SO_4**

Step 6: Use the number of moles of Na_2SO_4 from **step 5** and its molar mass to determine the **mass** produced:

Mass of 1 mol Na_2SO_4 = (2 × 23) + 32 + (4 × 16) g

= **142 g**

∴ mass of **0.3 mol Na_2SO_4** = 0.3 × 142 g

= **42.6 g**

Mass of sodium sulfate produced = **42.6 g**

Sample questions

1 What volume of hydrogen, measured at rtp, would be produced when 10.8 g of aluminium reacts with excess hydrochloric acid?

Steps 1 and **2**:

$$2Al(s) + 6HCl(aq) \longrightarrow 2AlCl_3(aq) + 3H_2(g)$$

10.8 g — **? volume at stp**

Mass of **Al** is known, volume of **H_2** is unknown.

Step 3: Mass of 1 mol Al = **27 g**

∴ number of moles in **10.8 g Al** = $\frac{10.8}{27}$ mol

= **0.4 mol**

Step 4: **2 mol Al** form **3 mol H_2**

Step 5: **0.4 mol Al** forms $\frac{3}{2}$ × 0.4 mol H_2

= **0.6 mol H_2**

Step 6: Volume of 1 mol H_2 at rtp = 24 dm^3

∴ volume of **0.6 mol H_2** = 0.6 × 24 dm^3

= **14.4 dm^3**

Volume of hydrogen produced = **14.4 dm^3**

2 What is the minimum mass of zinc oxide that must be added to 250 cm^3 of nitric acid with a concentration of 0.2 mol dm^{-3} to exactly neutralise the acid?

Steps 1 and **2:**

$$ZnO(s) + 2HNO_3(aq) \longrightarrow Zn(NO_3)_2(aq) + H_2O(l)$$

? mass   **250 cm^3** **0.2 mol dm^{-3}**

Volume and concentration of **HNO_3(aq)** are known, mass of **ZnO** is unknown.

Step 3: 1 dm^3 HNO_3(aq) contains 0.2 mol HNO_3

i.e. 1000 cm^3 HNO_3(aq) contains 0.2 mol HNO_3

$\therefore$ **250 cm^3 HNO_3(aq)** contains $\frac{0.2}{1000} \times 250$ mol HNO_3

$= \mathbf{0.05\ mol\ HNO_3}$

Step 4: **1 mol ZnO** reacts with **2 mol HNO_3**

Step 5: $\frac{1}{2} \times 0.05$ mol ZnO reacts with **0.05 mol HNO_3**

$= \mathbf{0.025\ mol\ ZnO}$

Step 6: Mass of 1 mol ZnO = 65 + 16 g = **81 g**

$\therefore$ mass of **0.025 mol ZnO** = 0.025 × 81 g

= **2.025 g**

Mass of zinc oxide required = **2.025 g**

3 What mass of copper(II) hydroxide would be produced when a solution containing 6.8 g of OH^- ions reacts with a solution containing excess Cu^{2+} ions?

Steps 1 and **2:**

$$Cu^{2+}(aq) + 2OH^-(aq) \longrightarrow Cu(OH)_2(s)$$

6.8 g   **? mass**

Mass of **OH^- ions** is known, mass of **$Cu(OH)_2$** is unknown.

Step 3: Mass of 1 mol OH^- ions = 16 + 1 g

= **17 g**

$\therefore$ number of moles in **6.8 g OH^- ions** = $\frac{6.8}{17}$ mol

= **0.4 mol**

Step 4: **2 mol OH^- ions** form **1 mol $Cu(OH)_2$**

Step 5: **0.4 mol OH^- ions** forms $\frac{1}{2} \times 0.4$ mol $Cu(OH)_2$

$= \mathbf{0.2\ mol\ Cu(OH)_2}$

Step 6: Mass of 1 mol $Cu(OH)_2$ = 64 + (2 × 16) + (2 × 1) g

= **98 g**

$\therefore$ mass of **0.2 mol $Cu(OH)_2$** = 0.2 × 98 g

= **19.6 g**

Mass of copper(II) hydroxide produced = **19.6 g**

Revision questions

9 State the Law of Conservation of Matter.

10 Potassium hydroxide reacts with phosphoric acid (H_3PO_4(aq)) to produce potassium phosphate and water. What mass of potassium phosphate could be produced when a solution containing 25.2 g of potassium hydroxide reacts with excess phosphoric acid?

11 Under the right conditions, hydrogen and oxygen react to form steam. What volume of oxygen would be required to react with excess hydrogen to produce 672 cm^3 of steam if the volumes are measured at stp?

12 Zinc reacts with hydrochloric acid (HCl(aq)) to form zinc chloride and hydrogen. What mass of zinc chloride could be produced if 25 cm^3 of hydrochloric acid of concentration 2.4 mol dm^{-3} reacts with excess zinc?

13 Sodium carbonate reacts with nitric acid (HNO_3(aq)) to produce sodium nitrate, carbon dioxide and water. What volume of carbon dioxide, measured at rtp, would be evolved if 50 cm^3 of nitric acid of concentration 2.0 mol dm^{-3} reacts with excess sodium carbonate?

14 When magnesium nitrate is heated the following reaction occurs:

$$2Mg(NO_3)_2(aq) \longrightarrow 2MgO(s) + 4NO_2(g) + O_2(g)$$

Determine the decrease in mass that would occur if 5.92 g of magnesium nitrate is heated until no further change occurs.

9 Acids, bases and salts

Acids and **bases** have **opposite** properties and have the ability to neutralise each other. When an acid reacts with a base, the reaction always forms a **salt**.

Acids

Acids which are not dissolved in water are composed of **covalent molecules** and can be solid, liquid or gas at room temperature. All acid molecules contain **hydrogen**, such as hydrochloric acid, HCl, nitric acid, HNO_3, and sulfuric acid, H_2SO_4. When added to water, the acid molecules **ionise**, forming positive **hydrogen ions (H^+ ions)** and negative anions:

e.g.

$$HCl(g) + water \longrightarrow H^+(aq) + Cl^-(aq)$$

$$H_2SO_4(l) + water \longrightarrow 2H^+(aq) + SO_4^{2-}(aq)$$

An acid may be defined in **two** ways:

- *An **acid** is a substance containing **hydrogen** which can be replaced directly or indirectly by a metal to form a salt.*

  e.g.

  $$Mg(s) + 2HCl(aq) \longrightarrow MgCl_2(aq) + H_2(g)$$

  The magnesium replaces the hydrogen to form a salt called magnesium chloride.

- *An **acid** is a **proton donor**.*

  H^+ ions, present in aqueous acids, are **single protons**. This is because each H^+ ion is a hydrogen nucleus containing a single proton which is formed by a hydrogen atom, ${}^1_1H$, losing its valence electron.

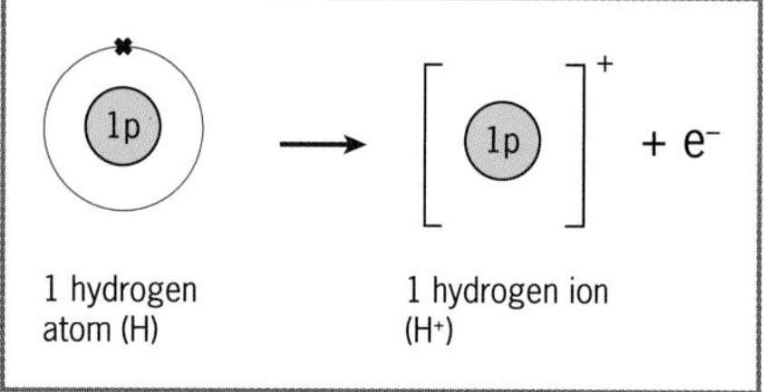

When an acid reacts it can **give** ('**donate**') its H^+ ions or protons to the other reactant. For example, when sodium carbonate reacts with hydrochloric acid, the acid **donates** its H^+ ions, or **protons**, to the CO_3^{2-} ions, forming carbon dioxide and water:

$$CO_3^{2-}(aq) + 2H^+(aq) \longrightarrow CO_2(g) + H_2O(l)$$

donated

General properties of acids in aqueous solution

The presence of **H^+ ions** in aqueous solutions of acids gives them their characteristic properties. These solutions are described as being **acidic** and they have the following properties:

- They have a **sour** taste.
- They are **corrosive**.
- They change blue litmus to **red**.
- They have a pH value of **less than 7**.
- They conduct an electric current, i.e. they are **electrolytes**.

Chemical reactions of acids in aqueous solution

When acids react, the **H^+ ions** in the acid are replaced by metal or ammonium ions to form a **salt**. Salts therefore contain metal or ammonium cations and negative anions from the acid. Aqueous acids undergo the following reactions:

- **Acids react with reactive metals**

  Acids, except nitric acid, react with metals above hydrogen in the reactivity series (see p. 158), to form a **salt** and **hydrogen**.

  reactive metal + acid ⟶ salt + hydrogen

  e.g. $Zn(s) + 2HCl(aq) \longrightarrow ZnCl_2(aq) + H_2(g)$

  Ionically: $Zn(s) + 2H^+(aq) \longrightarrow Zn^{2+}(aq) + H_2(g)$

  Note Nitric acid is an oxidising agent (see p. 86) which releases **oxides of nitrogen**, e.g. nitrogen dioxide (NO_2), and not hydrogen, when it reacts with metals.

- **Acids react with bases**

  Acids react with bases, which are mainly metal hydroxides and metal oxides (see p. 67), to form a **salt** and **water**.

  base + acid ⟶ salt + water

  e.g. $2KOH(aq) + H_2SO_4(aq) \longrightarrow K_2SO_4(aq) + 2H_2O(l)$

  Ionically: $OH^-(aq) + H^+(aq) \longrightarrow H_2O(l)$

  Or $CaO(s) + 2HCl(aq) \longrightarrow CaCl_2(aq) + H_2O(l)$

  Ionically: $CaO(s) + 2H^+(aq) \longrightarrow Ca^{2+}(aq) + H_2O(l)$

- **Acids react with metal carbonates and metal hydrogencarbonates**

  Acids react with metal carbonates and metal hydrogencarbonates to form a **salt**, **carbon dioxide** and **water**.

  metal carbonate + acid ⟶ salt + carbon dioxide + water

  e.g. $Na_2CO_3(aq) + H_2SO_4(aq) \longrightarrow Na_2SO_4(aq) + CO_2(g) + H_2O(l)$

  Ionically: $CO_3^{2-}(aq) + 2H^+(aq) \longrightarrow CO_2(g) + H_2O(l)$

  metal hydrogencarbonate + acid ⟶ salt + carbon dioxide + water

  e.g. $Mg(HCO_3)_2(aq) + 2HNO_3(aq) \longrightarrow Mg(NO_3)_2(aq) + 2CO_2(g) + 2H_2O(l)$

  Ionically: $HCO_3^-(aq) + H^+(aq) \longrightarrow CO_2(g) + H_2O(l)$

The basicity of acids

Acids can also be **classified** according to their **basicity**.

__Basicity__ is the number of __H^+ ions__ produced per molecule of acid when the acid dissolves in water.

- **Monobasic acids** produce **one** H^+ ion per molecule:

  e.g. $HCl(aq) \longrightarrow H^+(aq) + Cl^-(aq)$

  $HNO_3(aq) \longrightarrow H^+(aq) + NO_3^-(aq)$

  Monobasic acids can only form **normal salts** (see p. 71).

- **Dibasic acids** produce **two** H^+ ions per molecule:

  e.g. $$H_2SO_4(aq) \longrightarrow 2H^+(aq) + SO_4^{2-}(aq)$$

  Dibasic acids can form both **normal salts** and **acid salts** (see p. 71).
- **Tribasic acids** produce **three** H^+ ions per molecule:

  e.g. $$H_3PO_4(aq) \longrightarrow 3H^+(aq) + PO_4^{3-}(aq)$$

  Tribasic acids can form both **normal salts** and **acid salts**.

Acid anhydrides

*An **acid anhydride** is a compound that reacts with water to form an **acid**.*

Many acid anhydrides are **acidic oxides** of non-metals. Examples include carbon dioxide (CO_2), sulfur dioxide (SO_2), sulfur trioxide (SO_3) and nitrogen dioxide (NO_2).

$$CO_2(g) + H_2O(l) \rightleftharpoons \underset{\text{carbonic acid}}{H_2CO_3(aq)}$$

$$SO_2(g) + H_2O(l) \rightleftharpoons \underset{\text{sulfurous acid}}{H_2SO_3(aq)}$$

$$SO_3(g) + H_2O(l) \longrightarrow \underset{\text{sulfuric acid}}{H_2SO_4(aq)}$$

$$2NO_2(g) + H_2O(l) \longrightarrow \underset{\text{nitrous acid}}{HNO_2(aq)} + \underset{\text{nitric acid}}{HNO_3(aq)}$$

Acids in living systems

Acids can be found naturally in living organisms and some are used in everyday activities.

Table 9.1 *Acids in living organisms and everyday activities*

Acid	Where found	Notes
Ascorbic acid or vitamin C ($C_6H_8O_6$)	In many fruits and vegetables, e.g. West Indian cherries, citrus fruit, raw green vegetables.	• Vitamin C is essential in a healthy diet. A shortage can lead to **scurvy**. • When exposed to heat during cooking, vitamin C is **oxidised** which destroys it. • Sodium hydrogencarbonate is occasionally added to fruits and vegetables to improve appearance and texture. This **neutralises** any vitamin C in the foods which reduces its content.
Citric acid ($C_6H_8O_7$)	In citrus fruit, e.g. limes and lemons.	• Lime juice can be used to **remove rust stains** from clothing. The acid in the juice reacts with the iron(III) oxide (Fe_2O_3) in the rust stains. This makes a **soluble** compound which washes out of the clothes removing the rusty yellow Fe^{3+} ions: $\underset{\text{iron(III) oxide i.e. rust}}{Fe_2O_3(s)} + \underset{\text{in the acid}}{6H^+(aq)} \longrightarrow \underset{\text{wash out of clothes}}{2Fe^{3+}(aq)} + 3H_2O(l)$
Methanoic acid (HCOOH)	In the venom of ants.	• Methanoic acid can cause itching, swelling, redness and pain around the sting. • Ant stings can be treated by applying a paste of sodium hydrogencarbonate or calamine lotion which contains zinc oxide. Both compounds **neutralise** the acid.

Acid	Where found	Notes
Lactic acid ($C_3H_6O_3$)	Produced in the cells of muscles during strenuous activity.	• A person **collapses** if too much lactic acid builds up in the **muscles** because it prevents the muscles from contracting.
Ethanoic acid (CH_3COOH)	In vinegar.	• Vinegar can be used to **preserve** food items. Its low pH denatures (destroys) enzymes that cause decay and prevents the growth of microorganisms, i.e. bacteria and fungi.

Bases

Bases are chemically opposite to acids. Bases include **metal oxides**, e.g. calcium oxide (CaO), **metal hydroxides**, e.g. zinc hydroxide ($Zn(OH)_2$) and **ammonia** (NH_3).

*A **base** is a **proton acceptor**.*

When a **base** reacts with an **acid**, the O^{2-} ions or OH^- ions in the base **accept** the H^+ ions, or **protons**, from the acid, forming water. For example, when sodium hydroxide reacts with hydrochloric acid:

$$OH^-(aq) + H^+(aq) \longrightarrow H_2O(l)$$

accepted

Alkalis

*An **alkali** is a base which dissolves in water to form a solution that contains **OH^- ions**.*

Since most bases are **insoluble** in water, most bases are **not** alkalis.

Alkalis include:

- Potassium hydroxide (KOH) and sodium hydroxide (NaOH) which are soluble in water, and calcium hydroxide ($Ca(OH)_2$) which is slightly soluble:

  e.g. $$NaOH(s) + \text{water} \longrightarrow Na^+(aq) + OH^-(aq)$$

- Ammonia gas (NH_3), potassium oxide (K_2O), sodium oxide (Na_2O) and calcium oxide (CaO) which react with water to form a solution containing hydroxide ions:

  e.g. $$NH_3(g) + H_2O(l) \rightleftharpoons NH_4^+(aq) + OH^-(aq)$$

  $$Na_2O(s) + H_2O(l) \longrightarrow 2Na^+(aq) + 2OH^-(aq)$$

General properties of aqueous solutions of alkalis

The presence of **OH^- ions** in aqueous solutions of alkalis gives them their characteristic properties. These solutions are described as being **alkaline** and they have the following properties:

- They have a **bitter** taste.
- They are **corrosive**.
- They feel **soapy**.
- They change red litmus to **blue**.
- They have a pH value **greater than** 7.
- They conduct an electric current, i.e. they are **electrolytes**.

Chemical reactions of bases

- **Bases react with acids**

  Bases react with acids to produce a **salt** and **water** (see p. 65).

- **Bases react with ammonium salts**

  When heated, bases react with ammonium salts to produce a **salt, ammonia** and **water.**

$$\textbf{base + ammonium salt} \longrightarrow \textbf{salt + ammonia + water}$$

e.g.

$$Mg(OH)_2(s) + 2NH_4Cl(s) \longrightarrow MgCl_2(s) + 2NH_3(g) + 2H_2O(l)$$

$$CaO(s) + (NH_4)_2SO_4(s) \longrightarrow CaSO_4(s) + 2NH_3(g) + H_2O(l)$$

Distinguishing between acids and alkalis

Indicators are used to distinguish between **acids** and **alkalis** in aqueous solutions. An indicator has one colour in an acidic solution and another colour in an alkaline solution.

Table 9.2 *Common indicators*

Indicator	Colour in an acidic solution	Colour in an alkaline solution
Litmus	Red	Blue
Methyl orange	Red	Yellow
Screened methyl orange	Red	Green
Phenolphthalein	Colourless	Pink

The strength of acids and alkalis

The strength of an acid or alkali depends on the **degree of ionisation** which occurs when they dissolve in water.

- A **strong acid** is **fully ionised** when dissolved in water. All of the acid molecules ionise and the concentration of H^+ ions in the solution is **high.** Hydrochloric acid (HCl), sulfuric acid (H_2SO_4) and nitric acid (HNO_3) are strong acids.

  e.g. $$HCl(aq) \longrightarrow H^+(aq) + Cl^-(aq)$$

- A **weak acid** is only **partially ionised** when dissolved in water. The solution contains a mixture of acid molecules and H^+ ions, and the concentration of H^+ ions in the solution is **low**, e.g. ethanoic acid (CH_3COOH):

$$CH_3COOH(aq) \rightleftharpoons \underset{\text{ethanoate ion}}{CH_3COO^-(aq)} + H^+(aq)$$

- A **strong alkali** is **fully ionised** when dissolved in water. The concentration of OH^- ions in the solution is **high.** Potassium hydroxide (KOH) and sodium hydroxide (NaOH) are strong alkalis.

  e.g. $$NaOH(aq) \longrightarrow Na^+(aq) + OH^-(aq)$$

- A **weak alkali** is only **partially ionised** when dissolved in water. The concentration of OH^- ions in the solution is **low**, e.g. ammonia (NH_3):

$$NH_3(g) + H_2O(l) \rightleftharpoons NH_4^+(aq) + OH^-(aq)$$

Measuring the strength of acids and alkalis

The **strength** of an aqueous acid or alkali can be measured on the **pH scale** by using **universal indicator**.

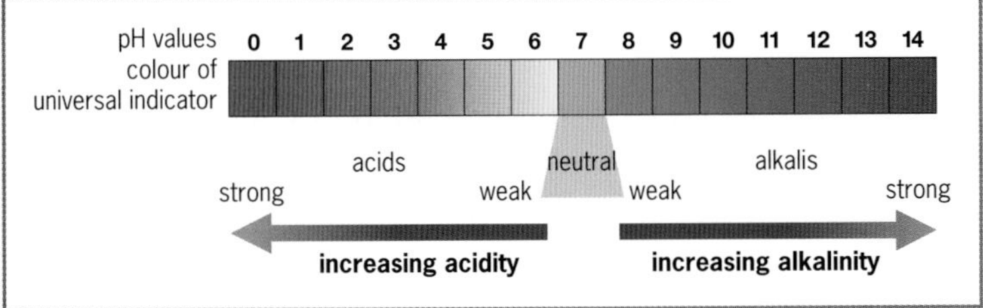

Figure 9.1 *The pH scale*

Note None of the indicators named in Table 9.2 measure pH.

Amphoteric oxides and hydroxides

*An **amphoteric oxide** or **hydroxide** can react with both **acids** and **strong alkalis** to from a salt and water.*

amphoteric oxide or hydroxide + acid ⟶ salt + water

strong alkali + amphoteric oxide or hydroxide ⟶ salt + water

The oxides and hydroxides of **aluminium**, **zinc** and **lead** are amphoteric.

Example

- **Zinc hydroxide** reacts with the acid, **hydrochloric acid:**

$$Zn(OH)_2(s) + 2HCl(aq) \longrightarrow ZnCl_2(aq) + 2H_2O(l)$$

- **Zinc hydroxide** reacts with the strong alkali, **sodium hydroxide**:

$$2NaOH(aq) + Zn(OH)_2(s) \longrightarrow \underset{\text{sodium zincate}}{Na_2ZnO_2(aq)} + 2H_2O(l)$$

Table 9.3 *Amphoteric oxides and hydroxides*

Amphoteric oxide	Amphoteric hydroxide	Salts formed when reacting with a strong alkali	Anion present in the salt
aluminium oxide (Al_2O_3)	aluminium hydroxide ($Al(OH)_3$)	**aluminates**	AlO_2^-
zinc oxide (ZnO)	zinc hydroxide ($Zn(OH)_2$)	**zincates**	ZnO_2^{2-}
lead(II) oxide (PbO)	lead(II) hydroxide ($Pb(OH)_2$)	**plumbates**	PbO_2^{2-}

Classification of oxides

Oxides can be classified into **four** groups:

- **Acidic oxides**

  Acidic oxides *are oxides of some* ***non-metals*** *which react with* ***alkalis*** *to form a salt and water.*

  Examples include carbon dioxide (CO_2), sulfur dioxide (SO_2), sulfur trioxide (SO_3), nitrogen dioxide (NO_2) and silicon dioxide (SiO_2).

  e.g. $2NaOH(aq) + CO_2(g) \longrightarrow Na_2CO_3(aq) + H_2O(l)$

  Most acidic oxides also react with **water** to form an acid, i.e. they are acid anhydrides (see p. 66).

- **Basic oxides**

  Basic oxides *are oxides of* ***metals*** *which react with* ***acids*** *to form a salt and water.*

  Examples include magnesium oxide (MgO), iron(III) oxide, (Fe_2O_3) and copper(II) oxide (CuO).

  e.g. $MgO(s) + H_2SO_4(aq) \longrightarrow MgSO_4(aq) + H_2O(l)$

  Potassium oxide (K_2O), sodium oxide (Na_2O) and calcium oxide (CaO) are basic oxides which are also classified as **alkalis** because they react with **water** to form a solution containing OH^- ions.

  e.g. $Na_2O(s) + H_2O(l) \longrightarrow 2NaOH(aq)$

- **Amphoteric oxides**

  Amphoteric oxides *are oxides of some* ***metals*** *which react with both* ***acids*** *and* ***strong alkalis*** *to form a salt and water.*

  There are three common amphoteric oxides, aluminium oxide (Al_2O_3), zinc oxide (ZnO) and lead(II) oxide (PbO) (see p. 69).

- **Neutral oxides**

  Neutral oxides *are oxides of some* ***non-metals*** *which do not react with acids or alkalis.*

  Examples include carbon monoxide (CO), nitrogen monoxide (NO) and dinitrogen monoxide (N_2O).

Revision questions

1 Explain why an acid can be defined as a proton donor.

2 Give FOUR properties, other than their chemical reactions, which are typical of aqueous acids.

3 Write balanced chemical equations for the reactions between:

a potassium carbonate and nitric acid

b sodium hydroxide and sulfuric acid

c calcium and hydrochloric acid

d magnesium hydrogencarbonate and sulfuric acid

4 Write ionic equations for EACH of the reactions in **3** above.

5 **a** What is an acid anhydride? **b** Name TWO acid anhydrides.

6 Explain the reason for EACH of the following:

a lime juice can be used to remove rust stains from clothes

b vinegar is used to preserve certain food items

7 Hydrochloric acid and ethanoic acid are both acidic; however, they have different pH values.

a Suggest a pH value for EACH acid.

b Explain the reason for the different pH values.

8 **a** What is a base?

b Explain the relationship between an alkali and a base.

9 Give FOUR properties, other than their chemical reactions, which are typical of aqueous alkalis.

10 Write balanced chemical equations for the reactions between:

a magnesium oxide and ammonium nitrate

b sodium hydroxide and ammonium sulfate

11 Identify the FOUR groups into which oxides can be classified, distinguish between them and give a named example of an oxide belonging to EACH group.

Salts

A ***salt*** *is a compound formed when some or all of the hydrogen ions in an acid are replaced by* ***metal*** *or* ***ammonium ions****.*

The metal or ammonium ions can come from the metal itself, a base, a carbonate or a hydrogencarbonate (see p. 65). Salts are **ionic compounds** which contain at least one **metal** or **ammonium cation** and one **anion** from the **acid**.

Classification of salts

Salts can be **classified** into **two** groups:

- **Normal salts** are formed when **all** of the H^+ ions in an acid are replaced by metal or ammonium ions.

  e.g. $$2KOH(aq) + H_2SO_4(aq) \longrightarrow K_2SO_4(aq) + 2H_2O(l)$$

  Potassium sulfate is a normal salt. **All** acids can form normal salts.

- **Acid salts** are formed when the H^+ ions in an acid are only **partially replaced** by metal or ammonium ions.

  e.g. $$KOH(aq) + H_2SO_4(aq) \longrightarrow KHSO_4(aq) + H_2O(l)$$

  Potassium hydrogensulfate is an acid salt. Only **dibasic** and **tribasic** acids can form acid salts.

The **relative quantity** of each reactant determines the **type** of salt formed by dibasic and tribasic acids. In the two reactions above:

- A **normal salt** is produced when **2 mol** of potassium hydroxide react with **1 mol** of sulfuric acid.
- An **acid salt** is produced when **1 mol** of potassium hydroxide reacts with **1 mol** of sulfuric acid.

Table 9.4 *Salts formed by some common acids*

Acid	Salts formed	Anion present	Type of salt	Name of the sodium salt	Formula
hydrochloric acid (HCl)	chlorides	Cl^-	normal	sodium chloride	NaCl
nitric acid (HNO_3)	nitrates	NO_3^-	normal	sodium nitrate	$NaNO_3$
ethanoic acid (CH_3COOH)	ethanoates	CH_3COO^-	normal	sodium ethanoate	CH_3COONa
sulfuric acid (H_2SO_4)	sulfates hydrogensulfates	SO_4^{2-} HSO_4^-	normal acid	sodium sulfate sodium hydrogensulfate	Na_2SO_4 $NaHSO_4$
carbonic acid (H_2CO_3)	carbonates hydrogencarbonates	CO_3^{2-} HCO_3^-	normal acid	sodium carbonate sodium hydrogencarbonate	Na_2CO_3 $NaHCO_3$
phosphoric acid (H_3PO_4)	phosphates hydrogenphosphates dihydrogenphosphates	PO_4^{3-} HPO_4^{2-} $H_2PO_4^-$	normal acid acid	sodium phosphate disodium hydrogenphosphate sodium dihydrogenphosphate	Na_3PO_4 Na_2HPO_4 NaH_2PO_4

Water of crystallisation

Some salts may contain **water of crystallisation**. This is a fixed proportion of water molecules held within their crystal lattice. Salts containing water of crystallisation are said to be **hydrated** and the water of crystallisation can be shown in the formula, e.g. $\mathbf{CuSO_4.5H_2O}$ represents hydrated copper(II) sulfate.

In some compounds, water of crystallisation is responsible for the **shape**, and sometimes the **colour**, of the crystals. If removed by heating, the salt becomes **anhydrous**; it loses its crystalline structure and its colour may change.

e.g. $CuSO_4.5H_2O(s) \xrightarrow{\text{heat}} CuSO_4(s) + 5H_2O(g)$

blue crystals – hydrated → white powder – anhydrous

A hydrated copper(II) sulfate crystal

Methods used to prepare salts

When preparing any salt, the following must be taken into account:

- The **solubility** of the **salt** being prepared.
- The **solubility** of the **reactants** being used to prepare the salt.
- The **hydration** of the **salt** being prepared.

The solubility of ionic compounds is summarised in Tables 6.1 and 6.2 on pp. 46 and 47.

Table 9.5 *Methods used to prepare* salts

Method of preparation	Salts prepared	Reactants used	Method	Example	Reactants and equation for the example
Ionic precipitation	**Insoluble salts**	**Two soluble salts:** one to supply the cations; one to supply the anions	• **Dissolve** each salt in distilled water to make two solutions. • **Mix** the solutions to form a precipitate. • **Filter** to separate the precipitate. • **Wash** the precipitate (residue) with distilled water. • **Dry** the residue.	**$PbSO_4$**	**$Pb(NO_3)_2(aq)$** to supply the **Pb^{2+} ions;** **$Na_2SO_4(aq)$** to supply the **SO_4^{2-} ions:** $Pb(NO_3)_2(aq) + Na_2SO_4(aq) \longrightarrow PbSO_4(s) + 2NaNO_3(aq)$ Ionically: $Pb^{2+}(aq) + SO_4^{2-}(aq) \longrightarrow PbSO_4(s)$
Titration	**Potassium,** **sodium** and **ammonium salts**	**Two solutions:** an **alkali** or **carbonate** solution to supply the K^+, Na^+ or NH_4^+ ions; a suitable **acid** to supply the anions	• Place the acid in a **burette** and find the volume needed to neutralise a **fixed volume** of alkali or carbonate solution, measured in a **pipette**, by performing a **titration** using a suitable **indicator**. • **Add** the volume of acid found above to the fixed volume of alkali or carbonate solution without adding the indicator to make a **normal salt**. • **Evaporate** the water, or evaporate some water and leave to **crystallise** (see p. 13).	**K_2SO_4** **$KHSO_4$**	**KOH(aq)** to supply the **K^+ ions;** **$H_2SO_4(aq)$** to supply the **SO_4^{2-} ions:** $2KOH(aq) + H_2SO_4(aq) \longrightarrow K_2SO_4(aq) + 2H_2O(l)$ [**1**] **KOH(aq)** to supply the **K^+ ions;** **$H_2SO_4(aq)$** to supply the **HSO_4^- ions:** $KOH(aq) + H_2SO_4(aq) \longrightarrow KHSO_4(aq) + H_2O(l)$ [**2**] Comparing equations [**1**] and [**2**], **twice** the volume of **acid** determined in the titration must be used to make the **acid salt.**

Salts in everyday life

Salts are important in everyday life, though some can be dangerous.

Table 9.6 *Uses of salts in everyday* life

Salt	Use	Notes
Sodium hydrogencarbonate ($NaHCO_3$)	A component of baking powder used to make cakes rise.	Baking powder also contains a weak acid and when mixed with the liquid in the cake mixture, the two active components react and form **carbon dioxide**: $HCO_3^-(aq) + H^+(aq) \longrightarrow CO_2(g) + H_2O(l)$ Carbon dioxide forms bubbles in the cake. These bubbles cause the cake to rise as they expand on heating.
Sodium benzoate (C_6H_5COONa)	To preserve food.	Used to preserve foods which have a **low pH**, e.g. fruit juices and fizzy drinks. At a low pH it is converted to benzoic acid which prevents the growth of microorganisms.
Sodium chloride ($NaCl$)	To preserve food.	Used to preserve food such as **meat** and **fish** (see p. 3).
Sodium nitrate ($NaNO_3$) and **sodium nitrite** ($NaNO_2$)	To preserve food.	Used to preserve **meat**, e.g. bacon and ham. They destroy bacteria which cause food poisoning, slow the oxidation of fats and oils which causes rancidity, give an attractive red colour to the meat and add flavour. They are often used together with sodium chloride.
Calcium carbonate (limestone) ($CaCO_3$)	To manufacture cement used in the construction industry.	When heated in a kiln, it decomposes to form **calcium oxide**: $CaCO_3(s) \longrightarrow CaO(s) + CO_2(g)$ Calcium oxide is blended with the other materials in the kiln to form **clinker** which is then ground with calcium sulfate to make cement.
Calcium sulfate (gypsum) ($CaSO_4.2H_2O$)	To manufacture plaster of Paris used as a building material and for setting broken bones.	Plaster of Paris is made of anhydrous calcium sulfate. When water is added, heat is given off and a **paste** forms. The paste is used to coat walls and ceilings, and bandages impregnated in it are used to make orthopaedic casts.
Magnesium sulfate (Epsom salt) ($MgSO_4.7H_2O$)	For various medicinal purposes.	Has numerous **health benefits**. Added to bath water it relieves stress, eases aches and pains, reduces inflammation and helps cure skin problems. Taken orally it works as a laxative.
	In agriculture.	Improves **plant growth**.

Table 9.7 *Dangers of* salts

Salt	Dangers of the salt
Sodium benzoate (C_6H_5COONa)	May increase the risk of developing **cancer** (may be **carcinogenic**). May increase **hyperactivity** and **asthma** in children.
Sodium chloride (NaCl)	Can lead to **hypertension** (high blood pressure) if consumed in excess.
Sodium nitrate ($NaNO_3$) and **sodium nitrite** ($NaNO_2$)	May increase the risk of developing **cancer** (may be **carcinogenic**). May cause **brain damage** in infants.

Neutralisation reactions

*A **neutralisation reaction** is a reaction between a base and an acid to form a salt and water.*

In a neutralisation reaction between an aqueous alkali and an aqueous acid, the **OH^- ions** of the alkali react with the **H^+ ions** of the acid:

$$OH^-(aq) + H^+(aq) \longrightarrow H_2O(l)$$

The **neutralisation point** or **end point** occurs when the OH^- ions have fully reacted with the H^+ ions and neither ion is present in excess. In a reaction between a **strong alkali** and a **strong acid**, the solution at this point is neutral, **pH 7**. Neutralisation reactions are **exothermic** since they produce heat energy.

To determine the neutralisation point in an acid-alkali reaction

The **neutralisation point** of a reaction between an aqueous alkali and an aqueous acid is determined by performing a **titration** using an **indicator** (see Table 9.2, p. 68) or **temperature change**.

- **Using an indicator**

  A fixed volume of alkali is measured using a pipette and run into a conical flask. A few drops of **indicator** are added and the acid is added from the burette. The neutralisation point is determined when the **colour** of the solution changes on the addition of a **single drop** of acid from the burette.

- **Using temperature change (a thermometric titration)**

  A fixed volume of alkali is placed into an insulated container and its temperature is recorded. The acid is added in small quantities, e.g. 2 cm^3, from a burette and the temperature is recorded after each addition until several successive drops in temperature have been recorded. A **graph** is drawn showing temperature against volume of acid added. Two straight lines of **best fit** are drawn and the **point of intersection** of the lines is the neutralisation point.

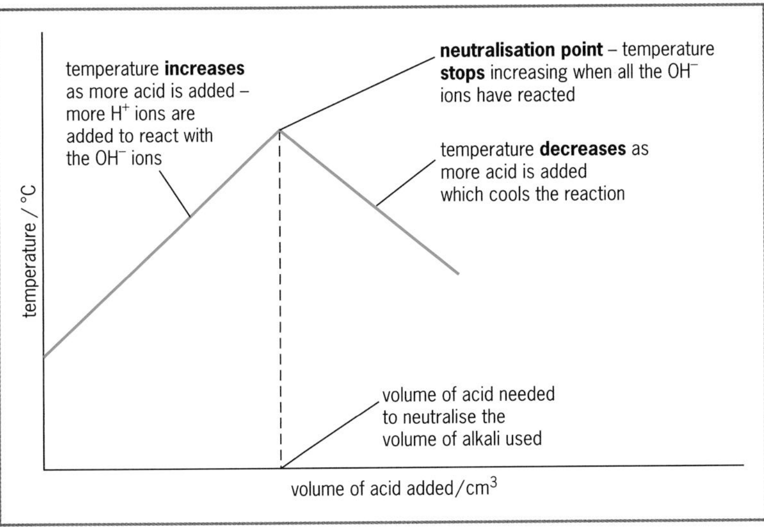

Figure 9.3 *Graph showing temperature against volume of acid added to a fixed volume of alkali*

Using neutralisation reactions in everyday life

Neutralisation reactions are used in our daily lives.

Table 9.8 *Neutralisation reactions in everyday* life

Use	Active compounds	Explanation
Toothpaste	Sodium hydrogencarbonate and sodium monofluorophosphate (Na_2FPO_3)	Toothpaste helps **reduce tooth decay** which is caused by **acid** in the mouth reacting with the calcium hydroxyapatite ($Ca_{10}(PO_4)_6(OH)_2$) in tooth enamel: • Sodium hydrogencarbonate **neutralises** any acid present. • F^- ions in the sodium monofluorophosphate **displace** the OH^- ions in the calcium hydroxyapatite forming calcium fluoroapatite ($Ca_{10}(PO_4)_6F_2$): $Ca_{10}(PO_4)_6(OH)_2(s) + 2F^-(aq) \longrightarrow Ca_{10}(PO_4)_6F_2(s) + 2OH^-(aq)$ Calcium fluoroapatite does not react with acid, therefore tooth enamel is protected from decaying.
Soil treatment	'Lime' in the form of calcium oxide (CaO) or calcium hydroxide ($Ca(OH)_2$)	Adding lime to soil **neutralises** any acids present, since most plants grow best if the soil pH is **neutral**. Lime cannot be added at the same time as an ammonium fertiliser because the two react to form a salt, ammonia and water: e.g. $CaO(s) + 2NH_4^+(aq) \longrightarrow Ca^{2+}(aq) + 2NH_3(g) + H_2O(l)$ The lime and the NH_4^+ ions are no longer available in the soil, so the benefits of both are lost.
Antacids	Sodium hydrogencarbonate ($NaHCO_3$), magnesium hydroxide ($Mg(OH)_2$), aluminium hydroxide ($Al(OH)_3$) or magnesium carbonate ($MgCO_3$)	Antacids are used to treat **indigestion** and **acid reflux**. They do this by **neutralising** excess hydrochloric acid in the stomach.

Volumetric analysis

Volumetric analysis involves performing a **titration** and using the results **quantitatively** in one of two ways:

- To calculate the **mole ratio** in which the two reactants combine.
- To calculate the **molar concentration** or **mass concentration** of one of the reactants used.

Using a titration to determine mole ratios

If **both** reactants are **standard solutions** (their mass or molar concentrations are known) the **mole ratio** in which they combine can be determined.

Example

To determine the mole ratio in which alkali *X* and acid *Y* combine, 25.0 cm^3 of alkali *X* of concentration 0.24 mol dm^{-3} was measured in a pipette and run into a conical flask. Acid *Y* of concentration 0.2 mol dm^{-3} was placed in a burette and a titration was performed. The results are given below:

	Titration number		
	Rough	**1**	**2**
Final burette reading/cm^3	15.1	30.1	45.1
Initial burette reading/cm^3	0.0	15.1	30.1
Volume of acid added/cm^3	15.1	15.0	15.0

Volume of acid *Y* needed to neutralise 25.0 cm^3 of alkali *X* = **15.0 cm^3**

Determine the **number of moles** of alkali *X* that reacted:

1 dm^3 of alkali *X*(aq) contains 0.24 mol *X*

i.e. 1000 cm^3 of alkali *X*(aq) contains 0.24 mol *X*

∴ **25.0 cm^3 of alkali *X*(aq)** contains $\frac{0.24}{1000} \times 25.0$ mol *X*

= **0.006 mol *X***

Determine the **number of moles** of acid *Y* that reacted:

1 dm^3 of acid *Y*(aq) contains 0.2 mol *Y*

i.e. 1000 cm^3 of acid *Y*(aq) contains 0.2 mol *Y*

∴ **15.0 cm^3 of acid *Y*(aq)** contains $\frac{0.2}{1000} \times 15.0$ mol *Y*

= **0.003 mol *Y***

Determine the **mole ratio** in which the reactants combine:

0.006 mol *X* reacts with 0.003 mol *Y*

∴ **2 mol *X*** react with **1 mol *Y***

Using a titration to determine concentration

If **one** reactant is a **standard solution** (its mass or molar concentration is known), the **concentration** of the other reactant can be determined by following the same **six** steps used on p. 60.

Example

To determine the molar concentration of a solution of sulfuric acid, 25.0 cm^3 of potassium hydroxide solution of concentration 0.4 mol dm^{-3} was measured in a pipette and run into a conical flask. The sulfuric acid was placed in a burette and a titration was performed. The results are given below:

	Titration number		
	Rough	**1**	**2**
Final burette reading/cm^3	20.1	40.3	20.0
Initial burette reading/cm^3	0.0	20.3	0.0
Volume of acid added/cm^3	20.1	20.0	20.0

Volume of sulfuric acid needed to neutralise 25.0 cm^3 of potassium hydroxide solution = **20.0 cm^3**

Steps 1 and **2:**

$$2KOH(aq) + H_2SO_4(aq) \longrightarrow K_2SO_4(aq) + 2H_2O(l)$$

2KOH(aq)	H_2SO_4(aq)
25.0 cm³	**20.0 cm³**
0.4 mol dm⁻³	**? concentration**

Volume and concentration of **KOH(aq)** are known, concentration of **H_2SO_4(aq)** is unknown.

Step 3: Find the **number of moles** of the known reactant, i.e. KOH, using its volume and concentration:

1 dm^3 KOH(aq) contains 0.4 mol KOH

i.e. 1000 cm^3 KOH(aq) contains 0.4 mol KOH

∴ **25.0 cm^3 KOH(aq)** contains $\frac{0.4}{1000} \times 25.0$ mol KOH

= **0.01 mol KOH**

Step 4: Use the balanced equation to determine the **mole ratio** between the KOH and the H_2SO_4:

2 mol KOH react with **1 mol H_2SO_4**

Step 5: Use the number of moles of KOH from **step 3** and the mole ratio from **step 4** to calculate the **number of moles** of H_2SO_4 that reacted:

0.01 mol KOH reacts with $\frac{1}{2} \times 0.01$ mol H_2SO_4

= **0.005 mol H_2SO_4**

Step 6: Use the number of moles of H_2SO_4 from **step 5** and the volume used in the titration to determine the **molar concentration** of the H_2SO_4(aq):

Since 20.0 cm^3 H_2SO_4 was used:

20.0 cm^3 H_2SO_4(aq) contains 0.005 mol H_2SO_4

∴ **1000 cm^3 H_2SO_4(aq)** contains $\frac{0.005}{20.0} \times 1000$ mol H_2SO_4

= **0.25 mol H_2SO_4**

Molar concentration of H_2SO_4(aq) = **0.25 mol dm^{-3}**

Sample question

During a titration it was found that 15.0 cm^3 of nitric acid of concentration 0.6 mol dm^{-3} neutralised 25.0 cm^3 of sodium carbonate solution of unknown concentration. Determine the mass concentration of the sodium carbonate solution.

Steps 1 and **2:**

$$Na_2CO_3(aq) + 2HNO_3(aq) \longrightarrow 2NaNO_3(aq) + H_2O(l)$$

Na_2CO_3(aq)	2HNO_3(aq)
25.0 cm³	**15.0 cm³**
? concentration	**0.6 mol dm⁻³**

Volume and concentration of **HNO_3(aq)** are known, concentration of **Na_2CO_3(aq)** is unknown.

Step 3: 1 dm^3 HNO_3(aq) contains 0.6 mol HNO_3

i.e. 1000 cm^3 HNO_3(aq) contains 0.6 mol HNO_3

∴ **15.0 cm^3 HNO_3(aq)** contains $\frac{0.6}{1000} \times 15.0$ mol HNO_3

= **0.009 mol HNO_3**

Step 4: **1 mol Na_2CO_3** reacts with **2 mol HNO_3**

Step 5: $\frac{1}{2} \times 0.009$ mol Na_2CO_3 reacts with **0.009 mol HNO_3**

= **0.0045 mol Na_2CO_3**

Step 6: Since 25.0 cm^3 Na_2CO_3 was used:

25.0 cm^3 $Na_2CO_3(aq)$ contains 0.0045 mol Na_2CO_3

∴ **1000 cm^3 $Na_2CO_3(aq)$** contains $\frac{0.0045}{25.0} \times 1000$ mol Na_2CO_3

= **0.18 mol Na_2CO_3**

i.e. molar concentration of $Na_2CO_3(aq)$ = **0.18 mol dm^{-3}**

Mass of 1 mol Na_2CO_3 = $(2 \times 23) + 12 + (3 \times 16)$ g

= **106 g**

∴ mass of **0.18 mol Na_2CO_3** = 0.18×106 g

= **19.08 g**

Mass concentration of $Na_2CO_3(aq)$ = **19.08 g dm^{-3}**

Revision questions

12 What is a salt?

13 Distinguish between a normal salt and an acid salt, and give a named example of EACH.

14 What is water of crystallisation?

15 Describe briefly, but including all essential experimental details and a relevant equation, how you would prepare a pure, dry sample of EACH of the following:

a zinc nitrate starting with zinc carbonate

b barium sulfate

c potassium sulfate starting with potassium hydroxide

d anhydrous iron(III) chloride

16 Name FOUR different salts used in daily life and give one use of each.

17 **a** What is a neutralisation reaction?

b What is the neutralisation point?

18 How does toothpaste help reduce tooth decay?

19 It was found that 7.5 cm^3 of hydrochloric acid of concentration 2.0 mol dm^{-3} neutralised 15.0 cm^3 of sodium carbonate solution of concentration 53.0 g dm^{-3}. Determine the mole ratio in which the reactants combined.

20 25.0 cm^3 of sulfuric acid of concentration 0.2 mol dm^{-3} exactly neutralised 40.0 cm^3 of sodium hydroxide solution. Determine the mass concentration of the sodium hydroxide solution.

10 Oxidation-reduction reactions

Oxidation and **reduction** are **opposite** processes that occur **together** in certain reactions. These are known as **redox reactions.**

Oxidation-reduction reactions in terms of electrons

In many redox reactions, one element **loses** electrons and another **gains** them. Oxidation and reduction can be defined in terms of electron transfer:

*Oxidation is the **loss** of electrons by an element in its free state, or an element in a compound.*

*Reduction is the **gain** of electrons by an element in its free state, or an element in a compound.*

To remember these definitions, remember two words, **OIL RIG**:

OIL	Oxidation	Is	Loss	RIG	Reduction	Is	Gain

Example

When aluminium is heated in a stream of chlorine gas aluminium chloride is formed:

$$2Al(s) + 3Cl_2(g) \longrightarrow 2AlCl_3(s)$$

Aluminium chloride ($AlCl_3$) is an ionic compound composed of $\mathbf{Al^{3+}}$ and $\mathbf{Cl^-}$ **ions.** In the reaction:

- Each aluminium atom **loses** three electrons to form an Al^{3+} ion:

  $$Al - 3e^- \longrightarrow Al^{3+}$$

  More correctly written as:

  $$Al \longrightarrow Al^{3+} + 3e^-$$

  Overall: $2Al(s) \longrightarrow 2Al^{3+}(s) + 6e^-$

  Aluminium (Al) has been **oxidised.**

- Each chlorine atom in each chlorine molecule **gains** one electron to form a Cl^- ion:

  $$Cl + e^- \longrightarrow Cl^-$$

  Overall: $3Cl_2(g) + 6e^- \longrightarrow 6Cl^-(s)$

  Chlorine (Cl_2) has been **reduced.**

Oxidation number

An **oxidation number** can be assigned to **each** atom or ion in a chemical substance. The oxidation number indicates the number of electrons lost, gained or shared as a result of chemical bonding. Oxidation numbers are either positive, negative or zero. Unless an oxidation number is zero, a **plus** or **minus sign** is written in front of the number, and the number 1 is always written, e.g. +1, –3.

Rules to follow when determining oxidation numbers

Rule 1: The oxidation number of each **atom** of an element in its free, uncombined state is **zero.**

e.g. In **Mg**, oxidation number of the **Mg atom = 0**

In $\mathbf{H_2}$, oxidation number of each **H atom = 0**

Rule 2: The oxidation number of each **monatomic ion** in an ionic compound is the same as the **charge** on the ion.

e.g. In $\mathbf{MgCl_2}$: oxidation number of the $\mathbf{Mg^{2+}}$ **ion** = **+2**

oxidation number of each $\mathbf{Cl^-}$ **ion** = **–1**

Rule 3: The oxidation number of **hydrogen** in a compound or polyatomic ion is always **+1**, except in metal hydrides where it is –1.

e.g. In $\mathbf{H_2O}$, oxidation number of **hydrogen** = **+1**

In the $\mathbf{NH_4^+}$ **ion**, oxidation number of **hydrogen = +1**

In $\mathbf{MgH_2}$ (magnesium hydride), oxidation number of **hydrogen = –1**

Rule 4: The oxidation number of **oxygen** in a compound or polyatomic ion is always **–2**, except in peroxides where it is –1.

e.g. In **MgO**, oxidation number of **oxygen = –2**

In the $\mathbf{SO_4^{2-}}$ **ion**, oxidation number of **oxygen = –2**

In $\mathbf{H_2O_2}$ (hydrogen peroxide), oxidation number of **oxygen = –1**

Rule 5: With the exception of hydrogen and oxygen, the oxidation numbers of elements in **covalent compounds** and **polyatomic ions** may **vary**. The oxidation number may appear in the **name** of the compound or ion:

e.g. In **sulfur(VI) oxide** (SO_3), oxidation number of **sulfur** = **+6**

In the **sulfate(IV) ion** (SO_3^{2-}), oxidation number of **sulfur** = **+4**

Rule 6: The **sum** of the oxidation numbers of all the atoms or ions in a **compound** is **zero**.

e.g. In $\mathbf{MgCl_2}$, the sum of the oxidation numbers of all the ions is **zero**:

(oxidation number of Mg) + 2(oxidation number of Cl) = 0

(+2) + 2(–1) = 0

(+2) + (–2) = 0

Rule 7: The **sum** of the oxidation numbers of all the atoms in a **polyatomic ion** is equal to the **charge on the ion.**

e.g. In the $\mathbf{OH^-}$ **ion**, the sum of the oxidation numbers of the two elements is **–1**:

(oxidation number of O) + (oxidation number of H) = –1

(–2) + (+1) = –1

If these rules are followed, it is possible to determine the oxidation number of any element from the **formula** of the compound or polyatomic ion it is in.

Examples

1. To determine the oxidation number of nitrogen in nitrogen dioxide (NO_2), rules **4** and **6** are applied:

(oxidation number of N) + 2(oxidation number of O) = 0

(oxidation number of N) + 2(–2) = 0

(oxidation number of N) + (–4) = 0

oxidation number of N = 0 + 4

= **+4**

Nitrogen dioxide can also be called **nitrogen(IV) oxide**.

2. To determine the oxidation number of nitrogen in the nitrite ion (NO_2^-), rules **4** and **7** are applied:

$$(\text{oxidation number of N}) + 2(\text{oxidation number of O}) = -1$$
$$(\text{oxidation number of N}) + 2(-2) = -1$$
$$(\text{oxidation number of N}) + (-4) = -1$$
$$\text{oxidation number of N} = -1 + 4$$
$$= \mathbf{+3}$$

The NO_2^- ion can be called the **nitrate(III) ion**.

Note When naming polyatomic ions using an oxidation number, the name always ends in '**-ate**'.

Sample questions

1 Determine the oxidation number of chromium in the $Cr_2O_7^{2-}$ ion and name the ion.

$$2(\text{oxidation number of Cr}) + 7(\text{oxidation number of O}) = -2$$
$$2(\text{oxidation number of Cr}) + 7(-2) = -2$$
$$2(\text{oxidation number of Cr}) + (-14) = -2$$
$$2(\text{oxidation number of Cr}) = -2 + 14$$
$$= +12$$
$$\text{oxidation number of Cr} = \frac{+12}{2}$$
$$= \mathbf{+6}$$

The oxidation number of chromium is **+6**. The ion is called the **dichromate(VI) ion**.

2 Determine the oxidation number of sulfur in sulfur dioxide (SO_2) and give an alternative name for sulfur dioxide.

$$S + 2(-2) = 0$$
$$S + (-4) = 0$$
$$S = \mathbf{+4}$$

The oxidation number of sulfur is **+4**. Sulfur dioxide can be called **sulfur(IV) oxide**.

3 Determine the oxidation number of carbon in ethane (C_2H_6).

$$2C + 6(+1) = 0$$
$$2C + 6 = 0$$
$$2C = -6$$
$$C = \frac{-6}{2}$$
$$= \mathbf{-3}$$

The oxidation number of carbon is **–3**.

Oxidation-reduction reactions in terms of oxidation number

In all redox reactions, the oxidation number of one element increases and the oxidation number of another element decreases. Oxidation and reduction can be defined in terms of oxidation number:

Oxidation *is the* ***increase*** *in oxidation number of an element in its free state, or an element in a compound.*

Reduction *is the* ***decrease*** *in oxidation number of an element in its free state, or an element in a compound.*

Example

The displacement reaction between chlorine and potassium bromide:

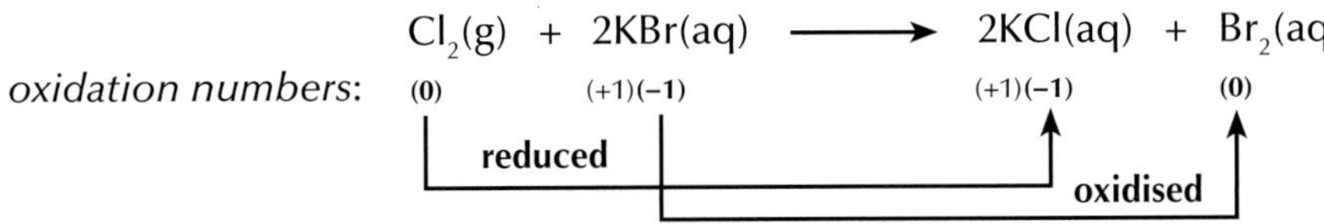

During the reaction:

- The oxidation number of each Br^- ion in the potassium bromide has **increased** from –1 to 0. **Potassium bromide (KBr)** has been **oxidised**.
- The oxidation number of each chlorine atom in the chlorine molecule has **decreased** from 0 to –1. **Chlorine (Cl_2)** has been **reduced**.

Using oxidation numbers to recognise redox reactions

Any redox reaction can be recognised using the following steps:

- Write the **balanced chemical equation** for the reaction if it has not been given.
- Write the **oxidation number** of each element below it in brackets. The oxidation numbers of elements in **polyatomic ions** which remain **unchanged** during a reaction need not be determined.
- Decide which element shows an **increase** in oxidation number. This element has been **oxidised**.
- Decide which element shows a **decrease** in oxidation number. This element has been **reduced**.

Note If the oxidation numbers of **all** elements remain **unchanged**, the reaction is **not** a redox reaction.

Sample questions

1 Determine which reactant has been oxidised and which has been reduced in the reaction between iron(III) oxide and carbon monoxide.

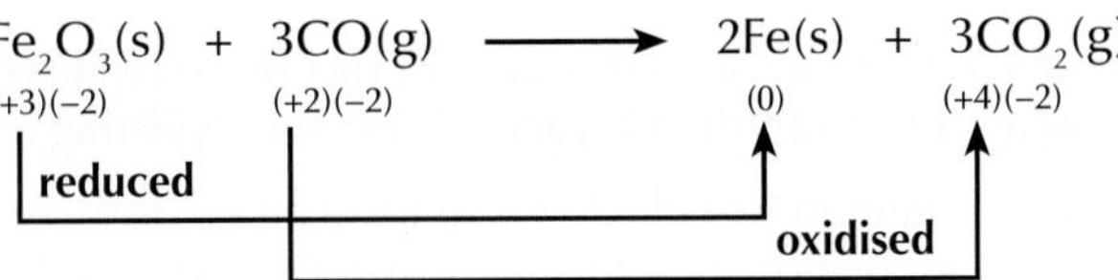

- **Carbon monoxide (CO)** has been **oxidised** because the oxidation number of each carbon atom in the carbon monoxide molecules has increased from +2 to +4.
- **Iron(III) oxide (Fe_2O_3)** has been **reduced** because the oxidation number of each Fe^{3+} ion in the iron(III) oxide has decreased from +3 to 0.

2 Determine which reactant has been oxidised and which has been reduced in the reaction between magnesium and sulfuric acid.

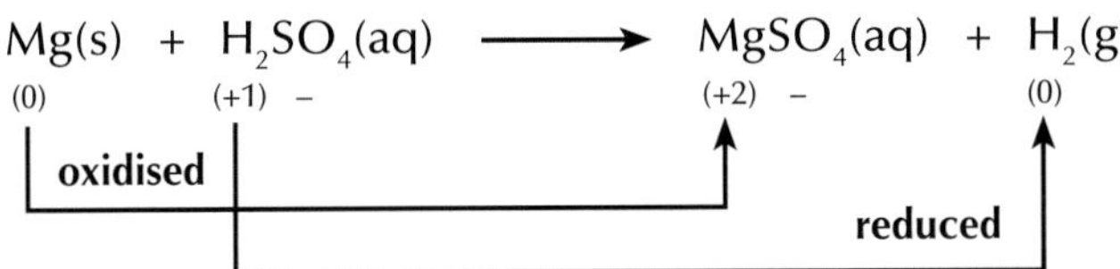

- **Magnesium (Mg)** has been **oxidised** because the oxidation number of the magnesium atom has increased from 0 to +2.
- **Sulfuric acid (H_2SO_4)** has been **reduced** because the oxidation number of each H^+ ion in the sulfuric acid has decreased from +1 to 0.

Oxidising and reducing agents

During any redox reaction:

- One reactant causes another reactant to be **oxidised.** This is the **oxidising agent.**
- One reactant causes another reactant to be **reduced.** This is the **reducing agent.**

In the following reaction, **X** has been **oxidised** and **Y** has been **reduced:**

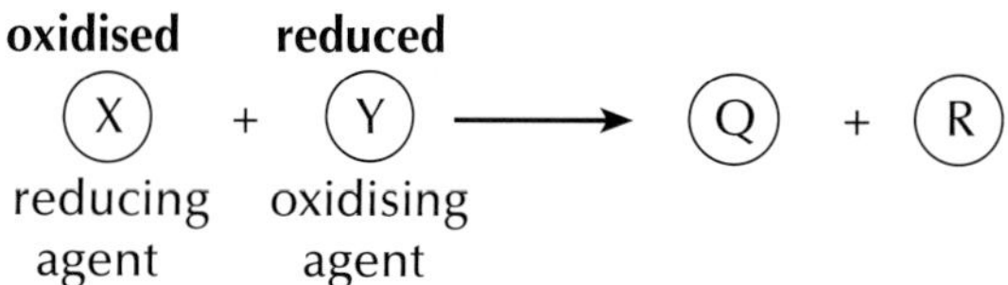

- **Y** must have caused X to be **oxidised. Y** is the **oxidising agent.**
- **X** must have caused Y to be **reduced. X** is the **reducing agent.**

Oxidising and reducing agents in terms of electrons

*An **oxidising agent** causes an element in its free state, or an element in a compound, to **lose electrons.***

*A **reducing agent** causes an element in its free state, or an element in a compound, to **gain electrons.***

Remember **OIL RIG.**

Example

The reaction between calcium and oxygen:

$$2Ca(s) + O_2(g) \longrightarrow 2CaO(s)$$

Where: $2Ca(s) \longrightarrow 2Ca^{2+}(s) + 4e^-$ (**oxidised**) and $O_2(g) + 4e^- \longrightarrow 2O^{2-}(s)$ (**reduced**)

- Each calcium atom lost two electrons to form a Ca^{2+} ion; calcium has been **oxidised.** The oxygen atoms caused this loss by taking away the electrons. **Oxygen (O_2)** is the **oxidising agent.**
- Each oxygen atom in the oxygen molecule gained two electrons to form an O^{2-} ion; oxygen has been **reduced.** The calcium atoms caused this gain by donating the electrons. **Calcium (Ca)** is the **reducing agent.**

Oxidising and reducing agents in terms of oxidation number

*An **oxidising agent** causes the oxidation number of an element in its free state, or an element in a compound, to **increase.***

*A **reducing agent** causes the oxidation number of an element in its free state, or an element in a compound, to **decrease.***

Example

$$Cl_2(g) + H_2S(g) \longrightarrow 2HCl(g) + S(s)$$

oxidation numbers: Cl_2 (0); H_2S (+1)(−2); HCl (+1)(−1); S (0)

Cl_2 (0) → HCl (−1): **reduced**; H_2S (−2) → S (0): **oxidised**

- The oxidation number of each sulfur atom in the hydrogen sulfide molecule has increased from −2 to 0. Hydrogen sulfide has been **oxidised.** Chlorine is the reactant that caused this increase in oxidation number. **Chlorine (Cl_2)** is the **oxidising agent.**
- The oxidation number of each chlorine atom in the chlorine molecule has decreased from 0 to −1. Chlorine has been **reduced.** Hydrogen sulfide is the reactant that caused this decrease in oxidation number. **Hydrogen sulfide (H_2S)** is the **reducing agent.**

Sample question

Determine which reactant is the oxidising agent and which is the reducing agent in the reaction between zinc and copper(II) sulfate.

$$\underset{(0)}{Zn(s)} + \underset{(+2)\ -}{CuSO_4(aq)} \longrightarrow \underset{(+2)\ -}{ZnSO_4(aq)} + \underset{(0)}{Cu(s)}$$

oxidised: Zn (0) → $ZnSO_4$ (+2); reduced: $CuSO_4$ (+2) → Cu (0)

- **Copper(II) sulfate ($CuSO_4$)** is the **oxidising agent** because it caused the oxidation number of the zinc atom to increase from 0 to +2.
- **Zinc (Zn)** is the **reducing agent** because it caused the oxidation number of the Cu^{2+} ion in the copper(II) sulfate to decrease from +2 to 0.

Common oxidising and reducing agents

Some substances always behave as **oxidising agents** and others always behave as **reducing agents**. A **visible change** may occur when some of these react.

- A **colour change** may occur.
- A **precipitate** may form.
- A particular **gas** may be produced.

Table 10.1 *Common oxidising agents*

Oxidising agent	Visible change when the agent reacts	Reason for the visible change
Acidified potassium manganate(VII) solution, **H^+/ $KMnO_4(aq)$**	Purple to **colourless.**	The purple MnO_4^- ion forms the colourless **Mn^{2+} ion.**
Acidified potassium dichromate(VI) solution, **H^+/ $K_2Cr_2O_7(aq)$**	Orange to **green.**	The orange $Cr_2O_7^{2-}$ ion forms the green **Cr^{3+} ion.**
Aqueous iron(III) salts, **$Fe^{3+}(aq)$**	Yellow-brown to **pale green**	The yellow-brown Fe^{3+} ion forms the pale green **Fe^{2+} ion.**
Sodium chlorate(I) solution, **NaClO(aq)**	Turns many coloured dyes **colourless** (see Table 10.3, p. 89).	The dyes are oxidised to their colourless form.
Hot concentrated sulfuric acid, **$H_2SO_4(l)$**	A **pungent** colourless gas is evolved.	**Sulfur dioxide** gas (SO_2) is produced.
Dilute or concentrated nitric acid, **$HNO_3(aq)$**	A **brown** gas is evolved.	**Nitrogen dioxide** gas (NO_2) is produced.

Table 10.2 *Common reducing agents*

Reducing agent	Visible change when the agent reacts	Reason for the visible change
Potassium iodide solution, **KI(aq)**	Colourless to **brown.**	**Iodine** (I_2) forms which dissolves forming a brown solution.
Aqueous iron(II) salts, **$Fe^{2+}(aq)$**	Pale green to **yellow-brown.**	The pale green Fe^{2+} ion forms the yellow-brown **Fe^{3+} ion.**
Hydrogen sulfide gas, **$H_2S(g)$**	A **yellow** precipitate forms.	Solid **sulfur** (S) forms.
Concentrated hydrochloric acid, **HCl(aq)**	A **yellow-green** gas is evolved.	**Chlorine** gas (Cl_2) is produced.

- Other common **oxidising agents** include oxygen, $\mathbf{O_2}$, chlorine, $\mathbf{Cl_2}$, and manganese(IV) oxide, $\mathbf{MnO_2}$.
- Other common **reducing agents** include hydrogen, $\mathbf{H_2}$, carbon, **C**, carbon monoxide, **CO**, and reactive metals.

Substances that can behave as both oxidising and reducing agents

Some compounds can act as both oxidising and reducing agents. Their behaviour depends on the other reactant.

Acidified hydrogen peroxide, H^+/H_2O_2

Acidified hydrogen peroxide is usually an **oxidising agent**. If it reacts with a stronger oxidising agent than itself, it acts as a **reducing agent**:

- With **potassium iodide** solution it acts as an **oxidising agent** and **oxidises** the iodide ions to iodine.
- With both **acidified potassium manganate(VII)** solution and **acidified potassium dichromate(VI)** solution, both stronger oxidising agents than itself, it acts as a **reducing agent**. The acidified hydrogen peroxide **reduces** the purple MnO_4^- ion to the colourless Mn^{2+} ion; and the orange $Cr_2O_7^{2-}$ ion to the green Cr^{3+} ion, respectively.

Sulfur dioxide, SO_2

Sulfur dioxide is usually a **reducing agent**. If it reacts with a stronger reducing agent than itself, it acts as an **oxidising agent**:

- With both **acidified potassium manganate(VII)** solution and **acidified potassium dichromate(VI)** solution it acts as a **reducing agent** causing the reactions described above.
- With **hydrogen sulfide**, a stronger reducing agent than itself, it acts as an **oxidising agent** and **oxidises** the hydrogen sulfide to yellow sulfur.

Tests for oxidising and reducing agents

Certain tests can be performed in the laboratory to determine if an unknown substance is an oxidising or reducing agent.

Tests for the presence of an oxidising agent

To test to see if a substance is an **oxidising agent**, add it to a known **reducing agent**, which gives a visible change when **oxidised**. The reducing agents usually used are **potassium iodide solution** or an aqueous solution of an **iron(II) salt**.

- An oxidising agent causes **potassium iodide** solution to change from colourless to **brown** because it **oxidises** the colourless I^- ion to **iodine** which dissolves forming a brown solution.
- An oxidising agent causes an aqueous solution of an **iron(II) salt**, e.g. iron(II) sulfate, to change from pale green to **yellow-brown** because it **oxidises** the pale green Fe^{2+} ion to the yellow-brown $\mathbf{Fe^{3+}}$ **ion**.

Tests for the presence of a reducing agent

To test to see if a substance is a **reducing agent**, add it to a known **oxidising agent**, which gives a visible change when it is **reduced**. The oxidising agents usually used are **acidified potassium manganate(VII)** solution or **acidified potassium dichromate(VI)** solution.

- A reducing agent causes **acidified potassium manganate(VII)** solution to change from purple to **colourless** because it **reduces** the purple MnO_4^- ion to the colourless $\mathbf{Mn^{2+}}$ **ion**.
- A reducing agent causes **acidified potassium dichromate(VI)** solution to change from orange to **green** because it **reduces** the orange $Cr_2O_7^{2-}$ ion to the green $\mathbf{Cr^{3+}}$ **ion**.

Purple acidified potassium manganate(VII) solution turns colourless

Orange acidified potassium dichromate(VI) solution turns green

Oxidation-reduction reactions in everyday activities

Oxidation-reduction reactions are encountered and used in everyday activities.

Table 10.3 *Oxidation and reduction in everyday activities*

Activity	Oxidation-reduction reaction
Bleaches	**Chlorine bleaches** containing sodium chlorate(I) (NaClO), and **oxygen bleaches** containing hydrogen peroxide (H_2O_2), remove coloured stains by **oxidising** the coloured chemicals, or dyes, in the stain to their colourless form: e.g. $ClO^-(aq)$ + coloured dye $\longrightarrow$ $Cl^-(aq)$ + colourless dye
Rusting	When oxygen and moisture come into contact with iron and its alloy, steel, the iron is **oxidised** to form hydrated iron(III) oxide ($Fe_2O_3.xH_2O$) commonly known as **rust**.
Browning of cut fruits and vegetables	When some fruits and vegetables are peeled or cut, e.g. apples, bananas and potatoes, enzymes in the cells on the cut surfaces are exposed to oxygen in the air. These enzymes **oxidise** certain chemicals in the cells to brown compounds called **melanins** which cause the cut surfaces to turn brown.
Preserving food	**Sodium sulfite** (Na_2SO_3) and **sulfur dioxide** (SO_2) are reducing agents used as **food preservatives**, e.g. to preserve dried fruit, fruit juices, wine and certain shellfish. They prevent spoilage of foods by **preventing oxidation**, e.g. they prevent oxidation of wine to vinegar and oxidation of vitamin C in fruits and fruit juices. They also prevent browning by **reducing** any melanins back to their colourless form.

Revision questions

1. Define EACH of the following in terms of electrons:

   a oxidation **b** reduction

   c oxidising agent **d** reducing agent

2. State, with a reason, whether EACH of the following equations shows oxidation or reduction:

   a $I_2(aq) + 2e^- \longrightarrow 2I^-(aq)$

   b $Cu^+(aq) \longrightarrow Cu^{2+}(aq) + e^-$

   c $2Br^-(aq) \longrightarrow Br_2(aq) + 2e^-$

3. Determine the oxidation number of:

   a bromine in: bromine dioxide (BrO_2), the BrO_3^- ion, and the BrO^- ion

   b nitrogen in: ammonia (NH_3), the NO_2^- ion, and dinitrogen monoxide (N_2O)

   c carbon in: carbon monoxide (CO), the CO_3^{2-} ion, methane (CH_4), propene (C_3H_6), and the HCO_3^- ion.

4. **a** Determine the oxidation number of sulfur in the SO_3^{2-} ion and the SO_4^{2-} ion and use this number to name EACH ion.

   b Suggest an alternative name for nitrogen monoxide (NO) using oxidation number.

5. Define EACH of the following in terms of oxidation number:

   a oxidation **b** reduction

   c oxidising agent **d** reducing agent

6. State, with reasons based on oxidation number, which reactant has been oxidised and which has been reduced in EACH of the following reactions:

   a $2FeCl_3(aq) + H_2S(g) \longrightarrow 2FeCl_2(aq) + S(s) + 2HCl(aq)$

   b $CH_4(g) + 4CuO(s) \longrightarrow 4Cu(s) + CO_2(g) + 2H_2O(g)$

7. State, with reasons based on oxidation number, which reactant is the oxidising agent and which is the reducing agent in the following reaction:

   $Fe_2O_3(s) + 3H_2(g) \longrightarrow 2Fe(s) + 3H_2O(l)$

8. Name TWO reagents you could use to prove that an unknown substance is a reducing agent. For EACH reagent, give the colour change you would expect and explain this colour change.

9. Name TWO substances that can behave as both oxidising and reducing agents.

11 Electrochemistry

Electrochemistry is the study of **electrochemical reactions.** These are reactions that either produce electrical energy or require electrical energy to proceed.

Predicting reactions using the electrochemical series of metals

The **electrochemical series of metals** places metals in order of how easily they **lose** electrons (**ionise**), and can be used to **predict** certain chemical reactions. The ability of metal atoms give away (donate) electrons to another reactant **increases** going **up** the series. Therefore, the **strength as a reducing agent** increases going up the series.

Table 11.1 *The electrochemical series of the common metals*

Metal	Ease of ionisation	Cation
potassium	↑	K^+
calcium		Ca^{2+}
sodium		Na^+
magnesium		Mg^{2+}
aluminium		Al^{3+}
zinc		Zn^{2+}
iron		Fe^{2+}
lead		Pb^{2+}
hydrogen		H^+
copper		Cu^{2+}
silver		Ag^+

⟶ Indicates an **increase**

Displacement of metals

A metal displaces a metal **below** it in the series from a compound containing ions of the lower metal. The **higher** metal is a **stronger reducing agent**, so it readily **gives** electrons to the ions of the lower metal. In doing so, the higher metal **ionises** to form cations. The ions of the lower metal **gain** these electrons and are **discharged** to form atoms.

Example

$$Zn(s) + CuSO_4(aq) \longrightarrow ZnSO_4(aq) + Cu(s)$$

Zinc is above copper in the series. The zinc atoms **ionise** and form Zn^{2+} ions. The Cu^{2+} ions are **discharged** and form copper atoms:

$$Zn(s) \longrightarrow Zn^{2+}(aq) + 2e^-$$

$$Cu^{2+}(aq) + 2e^- \longrightarrow Cu(s)$$

Displacement of hydrogen

Metals **above** hydrogen in the series displace the H^+ ions in an **acid**, forming hydrogen gas. Metals **below** hydrogen do not react with acids because they do not displace the H^+ ions.

Example

$$Mg(s) + 2HCl(aq) \longrightarrow MgCl_2(aq) + H_2(g)$$

Magnesium is above hydrogen in the series. The magnesium atoms **ionise** and form Mg^{2+} ions. The H^+ ions are **discharged** and form hydrogen gas.

$$Mg(s) \longrightarrow Mg^{2+}(aq) + 2e^-$$

$$2H^+(aq) + 2e^- \longrightarrow H_2(g)$$

Predicting reactions using the electrochemical series of non-metals

The **electrochemical series of non-metals** places non-metals in order of how easily they **gain** electrons, (**ionise**), and can be used to **predict** certain chemical reactions. The ability of non-metal atoms to ionise and take away electrons from another reactant **increases** going **up** the series. Therefore, the **strength as an oxidising agent** increases going up the series.

Table 11.2 *The electrochemical series of certain non-metals*

Non-metal	Ease of ionisation	Anion
fluorine	↑	F^-
chlorine		Cl^-
bromine		Br^-
iodine		I^-

Displacement of non-metals

A non-metal displaces a non-metal **below** it in the series from a compound containing ions of the lower non-metal. The **higher** non-metal is a **stronger oxidising agent**, so it readily **takes** electrons from the ions of the lower non-metal. In doing so, the higher non-metal **ionises** to form anions. The ions of the lower non-metal **lose** these electrons and are **discharged** to form atoms.

Example

$$Cl_2(g) + 2KI(aq) \longrightarrow 2KCl(aq) + I_2(aq)$$

Chlorine is above iodine in the series. The chlorine atoms **ionise** and form Cl^- ions. The I^- ions are **discharged** to form iodine:

$$Cl_2(g) + 2e^- \longrightarrow 2Cl^-(aq)$$

$$2I^-(aq) \longrightarrow I_2(aq) + 2e^-$$

Electrical conduction

Based on their ability to conduct an electric current, materials can be classified into **two** groups:

- **Conductors** allow an electric current to pass through. Metals, graphite, molten ionic compounds, aqueous solutions of ionic compounds, acids and alkalis are conductors.
- **Non-conductors** do not allow an electric current to pass through. Non-metals (except graphite), plastics, solid ionic compounds, covalent compounds and aqueous solutions of covalent compounds are non-conductors.

Electrolytes

When an ionic compound melts or dissolves in water, the liquid or solution that forms is known as an **electrolyte**. Because the ionic bonds have broken and the ions are free to move, electrolytes are **conductors**. When an electric current passes through an electrolyte it **decomposes**.

Differences between metallic and electrolytic conduction

Differences exist between conduction in a metal (**metallic conduction**) and conduction in an electrolyte (**electrolytic conduction**).

Table 11.3 *Metallic and electrolytic conduction compared*

Conduction in a metal	Conduction in an electrolyte
Mobile electrons in the electron pool carry the electric current through the metal.	**Mobile ions**, which are no longer held together by ionic bonds, carry the electric current through the electrolyte.
The metal remains **unchanged** chemically.	The electrolyte **decomposes**, i.e. it is chemically changed.

Strength of electrolytes

The **strength** of an electrolyte is determined by the **concentration of ions** present:

- **Strong electrolytes** are **fully ionised** when they dissolve in water. Their solutions contain a **high** concentration of ions. Examples include aqueous solutions of ionic compounds, strong acids and strong alkalis.

  e.g. $$H_2SO_4(aq) \longrightarrow 2H^+(aq) + SO_4^{2-}(aq)$$

  Molten ionic compounds are also strong electrolytes.
- **Weak electrolytes** are only **partially ionised** when they dissolve in water. Their solutions contain a **low** concentration of ions. Examples include weak acids and weak alkalis.

  e.g. $$CH_3COOH(aq) \rightleftharpoons CH_3COO^-(aq) + H^+(aq)$$

Pure water

Pure water is an **extremely weak** electrolyte. Approximately one in every 5.56×10^8 water molecules is ionised into H^+ and OH^- ions at any one time:

$$H_2O(l) \rightleftharpoons H^+(aq) + OH^-(aq)$$

The presence of these ions is important when an electric current passes through an aqueous electrolyte (see p. 95).

Non-electrolytes

Non-electrolytes are substances which remain as molecules in the liquid state or dissolved in water, and therefore they do not contain any ions. Non-electrolytes are **non-conductors** and include:

- Liquids, e.g. kerosene, gasoline or ethanol.
- Molten covalent substances, e.g. molten wax.
- Solutions of covalent substances, e.g. solutions of glucose or ethanol.

Electrolysis

*Electrolysis is the **chemical change** occurring when an electric current passes through an **electrolyte**.*

Electrolysis is carried out in an **electrolytic cell**, which has **three** main components:

- The **electrolyte**. This is a molten ionic compound or solution which contains **mobile ions**.

- A **battery** or other DC power supply. This supplies the **electric current**.
- Two **electrodes**. These are connected to the power supply by wires and are placed in the electrolyte so that they can carry the current **into** and **out of** the electrolyte. They are usually made of an inert material, e.g. graphite (carbon) or platinum, which can conduct electricity:

  *The **anode** is the **positive** electrode, connected to the positive terminal of the power supply.*

  *The **cathode** is the **negative** electrode, connected to the negative terminal of the power supply.*

Principles of electrolysis

During electrolysis:

- The negative **anions** are attracted to the positive **anode** where they are **discharged** to form atoms by losing electrons to the anode:

$$N^{n-} \longrightarrow N + ne^-$$

Oxidation occurs at the **anode** (**O**xidation **I**s **L**oss, OIL). The anode behaves as the **oxidising agent**.

- The **electrons** which the anions lose at the anode move through the circuit from the anode to the positive terminal of the battery. The electrons re-enter the circuit from the negative terminal of the battery and move from the battery to the cathode.
- The positive **cations** are attracted to the negative **cathode** where they are **discharged** to form atoms by gaining electrons from the cathode:

$$M^{n+} + ne^- \longrightarrow M$$

Reduction occurs at the **cathode** (**R**eduction **I**s **G**ain, RIG). The cathode behaves as the **reducing agent**.

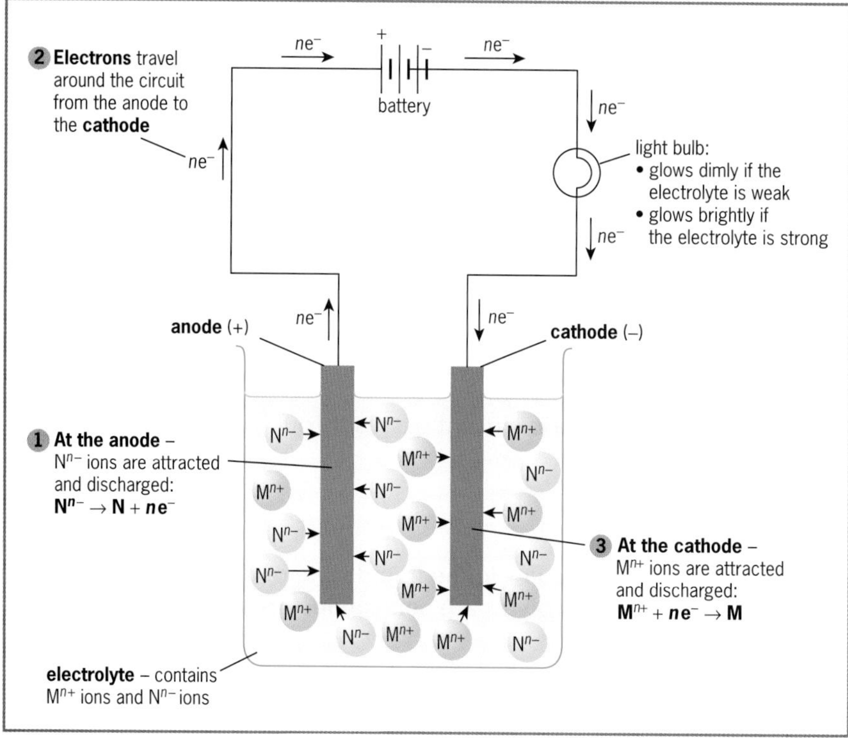

Figure 11.1 *Events occurring during electrolysis*

Electrolysis of molten electrolytes

Molten electrolytes contain **two** different ions only, a cation and an anion. Both ions are **discharged** during electrolysis.

Example

Electrolysis of molten (fused) lead(II) bromide using inert graphite electrodes

- **Ions present in the electrolyte:**

  $Pb^{2+}(l)$ and $Br^{-}(l)$

- **Events at the anode:**

  The **Br^{-} ions** move towards the anode where each ion loses an electron to the anode to form a bromine atom (the ions are discharged). Bromine atoms immediately bond covalently in pairs to form **bromine molecules:**

  $$2Br^{-}(l) \longrightarrow Br_2(g) + 2e^{-}$$

  Red-brown **bromine** vapour is evolved around the anode.

- **Events at the cathode:**

  The **Pb^{2+} ions** move towards the cathode where each ion gains two electrons to form a lead atom (the ions are discharged):

  $$Pb^{2+}(l) + 2e^{-} \longrightarrow Pb(l)$$

  Molten **lead** is formed around the cathode and drips off.

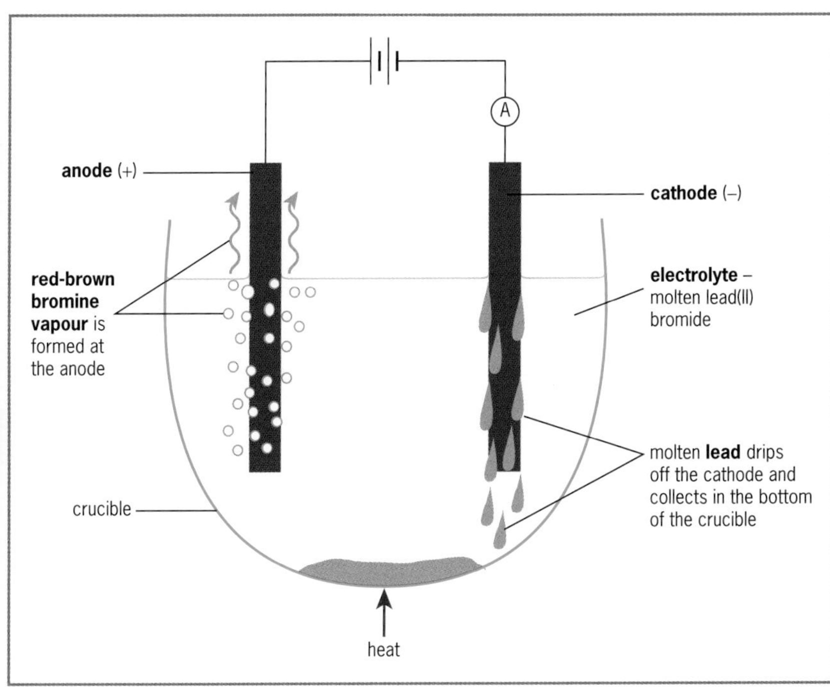

Figure 11.2 *Electrolysis of molten lead(II) bromide*

Electrolysis of aqueous electrolytes

An **aqueous solution** contains at least **two** different **cations** and **two** different **anions** because it contains ions from the **solute** and **H^{+} ions** and **OH^{-} ions** from the ionisation of water molecules. During electrolysis, **one** type of cation and **one** type of anion are discharged in preference to any others present. This is called **preferential discharge**.

Factors influencing the preferential discharge of anions

Three main factors influence the preferential discharge of the **anions**:

- **The type of anode**

  An anode which is **not inert**, e.g. copper, can take part in the electrolysis process and this affects what happens at the anode. If an **active** anode is used, the reaction occurring is the one which requires the least energy. This usually involves the anode **ionising** instead of an anion being discharged.

  Comparing the electrolysis of copper(II) sulfate solution using an inert anode and an active copper anode demonstrates this (see Table 11.6, p. 98).

- **The concentration of the electrolyte**

  The **greater** the concentration of an ion, the **more likely** it is to be preferentially discharged. This rule applies mainly to solutions containing **halide ions** (**Cl^-**, **Br^-** and **I^- ions**).

  Comparing the electrolysis of dilute and concentrated sodium chloride solutions using inert electrodes demonstrates this (see Table 11.6).

- **The position of the ion in the electrochemical series**

  The **lower** the ion in the electrochemical series of anions, the **more likely** it is to be preferentially discharged. Ions at the **top** of series are the **hardest to discharge** because they are the most stable. Ions at the **bottom** are the **easiest to discharge** because they are the least stable.

Table 11.4 *The electrochemical series of anions*

Anion
F^-
SO_4^{2-}
NO_3^-
Cl^-
Br^-
I^-
OH^-

The electrolysis of dilute sulfuric acid, dilute sodium chloride solution and copper(II) sulfate solution using inert electrodes all demonstrate this (see Table 11.6).

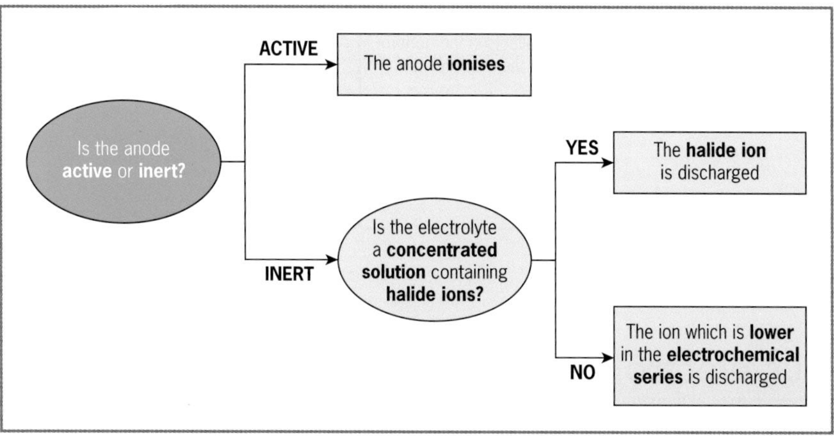

Figure 11.3 *A flow diagram to determine the reaction occurring at the anode*

Preferential discharge of cations

The **position** of the ion in the **electrochemical series** influences the **preferential discharge** of the cations.

The **lower** the ion in the electrochemical series of cations, the **more likely** it is to be preferentially discharged. Ions at the **top** of series are the **hardest to discharge** because they are the most stable. Ions at the **bottom** are the **easiest to discharge** because they are the least stable.

Table 11.5 *The electrochemical series of cations*

Cation
K^+
Ca^{2+}
Na^+
Mg^{2+}
Al^{3+}
Zn^{2+}
Fe^{2+}
Pb^{2+}
H^+
Cu^{2+}
Ag^+

The electrolysis of both dilute and concentrated sodium chloride solutions, and copper(II) sulfate solution demonstrate this (see Table 11.6).

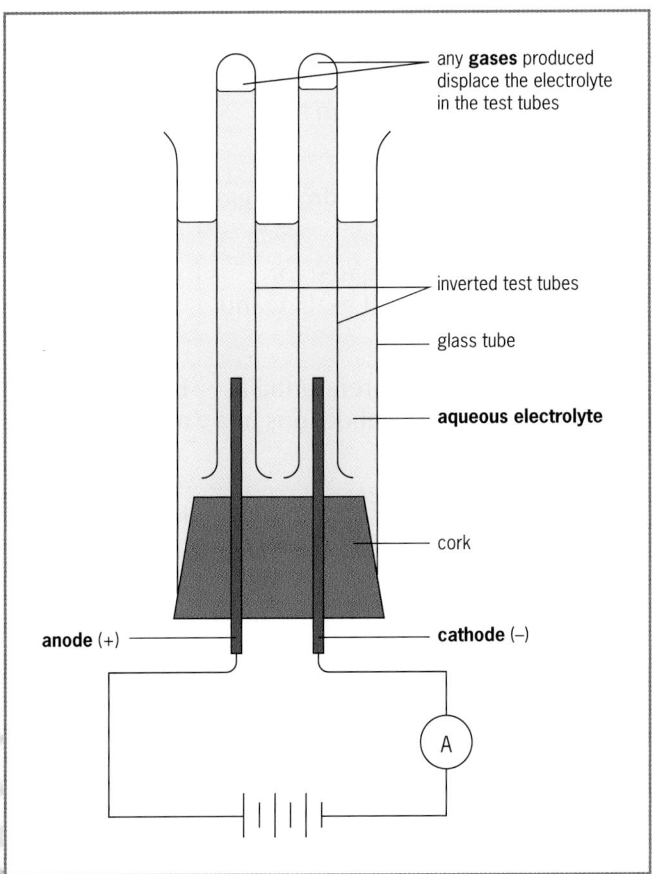

Figure 11.4 *An electrolytic cell used to electrolyse aqueous solutions*

Table 11.6 *The electrolysis of some aqueous solutions*

Electrolyte and ions present	Electrodes	Reactions at the electrodes and nature of the products	How the electrolyte changes
Dilute sulfuric acid, $H_2SO_4(aq)$ From H_2SO_4: $H^+(aq)$, $SO_4^{2-}(aq)$ From H_2O: $H^+(aq)$, $OH^-(aq)$	**Inert** – carbon or platinum	• **At anode: OH^- ions** are preferentially discharged, they are lower in the electrochemical series than SO_4^{2-}: $4OH^-(aq) \longrightarrow 2H_2O(l) + O_2(g) + 4e^-$ **Effervescence** occurs as **oxygen** gas is evolved. • **At cathode: H^+ ions** are discharged: $2H^+(aq) + 2e^- \longrightarrow H_2(g)$ **Effervescence** occurs as **hydrogen** gas is evolved. • **Relative proportions of gases**: for every 4 mol electrons, 1 mol O_2 and 2 mol H_2 are produced.	Becomes **more concentrated:** H^+ and OH^- ions are removed leaving H^+ and SO_4^{2-} ions in excess, i.e. water is removed.
Dilute sodium chloride solution, NaCl(aq) From NaCl: $Na^+(aq)$, $Cl^-(aq)$ From H_2O: $H^+(aq)$, $OH^-(aq)$	**Inert** – carbon or platinum	• **At anode: OH^- ions** are preferentially discharged, they are lower in the electrochemical series than Cl^-: $4OH^-(aq) \longrightarrow 2H_2O(l) + O_2(g) + 4e^-$ **Effervescence** occurs as **oxygen** gas is evolved. • **At cathode: H^+ ions** are preferentially discharged, they are lower in the electrochemical series than Na^+: $2H^+(aq) + 2e^- \longrightarrow H_2(g)$ **Effervescence** occurs as **hydrogen** gas is evolved. • **Relative proportions of gases**: for every 4 mol electrons, 1 mol O_2 and 2 mol H_2 are produced.	Becomes **more concentrated:** H^+ and OH^- ions are removed leaving Na^+ and Cl^- ions in excess, i.e. water is removed.
Concentrated sodium chloride solution, NaCl(aq) From NaCl: $Na^+(aq)$, $Cl^-(aq)$ From H_2O: $H^+(aq)$, $OH^-(aq)$	**Inert** – carbon or platinum	• **At anode: Cl^- ions** are preferentially discharged, they are halide ions in a concentrated solution: $2Cl^-(aq) \longrightarrow Cl_2(g) + 2e^-$ **Effervescence** occurs as **chlorine** gas is evolved. • **At cathode: H^+ ions** are preferentially discharged, they are lower in the electrochemical series than Na^+: $2H^+(aq) + 2e^- \longrightarrow H_2(g)$ **Effervescence** occurs as **hydrogen** gas is evolved. • **Relative proportions of gases**: for every 2 mol electrons, 1 mol Cl_2 and 1 mol H_2 are produced.	Becomes **alkaline:** H^+ and Cl^- ions are removed leaving Na^+ and OH^- ions in excess, i.e. sodium hydroxide is formed.

Electrolyte and ions present	Electrodes	Reactions at the electrodes and nature of the products	How the electrolyte changes
Copper(II) sulfate solution, $CuSO_4$(aq) From $CuSO_4$: Cu^{2+}(aq), SO_4^{2-}(aq) From H_2O: H^+(aq), OH^-(aq)	**Inert** – carbon or platinum	• **At anode: OH^- ions** are preferentially discharged, they are lower in the electrochemical series than SO_4^{2-}: $4OH^-(aq) \longrightarrow 2H_2O(l) + O_2(g) + 4e^-$ **Effervescence** occurs as **oxygen** gas is evolved. • **At cathode: Cu^{2+} ions** are preferentially discharged, they are lower in the electrochemical series than H^+: $Cu^{2+}(aq) + 2e^- \longrightarrow Cu(s)$ **Pink** copper is deposited and the cathode **increases** in size.	Becomes **acidic**: Cu^{2+} and OH^- ions are removed leaving H^+ and SO_4^{2-} ions in excess, i.e. sulfuric acid is formed. Becomes **paler blue**: blue Cu^{2+} ions are removed.
Copper(II) sulfate solution, $CuSO_4$(aq) From $CuSO_4$: Cu^{2+}(aq), SO_4^{2-}(aq) From H_2O: H^+(aq), OH^-(aq)	Anode – **active copper** Cathode – carbon, platinum or copper	• **At anode**: the anode **ionises**, this requires less energy than discharging either anion: $Cu(s) \longrightarrow Cu^{2+}(aq) + 2e^-$ Cu^{2+} ions go into the electrolyte and the anode **decreases** in size. • **At cathode: Cu^{2+} ions** are preferentially discharged, they are lower in the electrochemical series than H^+: $Cu^{2+}(aq) + 2e^- \longrightarrow Cu(s)$ **Pink** copper is deposited and the cathode **increases** in size.	Remains **unchanged**: for every 2 mol electrons, 1 mol Cu^{2+} ions enter the electrolyte at the anode, and 1 mol Cu^{2+} ions is discharged at the cathode.

Revision questions

1 Predict whether a displacement reaction will occur when EACH of the following are mixed and write a balanced equation for EACH reaction that you predict will occur.

a zinc and hydrochloric acid
b iron and zinc chloride solution
c aluminium and silver nitrate solution
d copper and sulfuric acid
e iodine and potassium chloride solution
f bromine and sodium iodide solution

2 Distinguish between a conductor and a non-conductor and give TWO named examples of EACH.

3 **a** What is an electrolyte?

b Give TWO differences between metallic conduction and electrolytic conduction.

4 Distinguish between a strong electrolyte and a weak electrolyte, and give a named example of EACH.

5 Define EACH of the following:

a electrolysis **b** anion **c** cation **d** anode **e** cathode

6 What THREE factors influence the preferential discharge of anions from aqueous solutions?

7 Describe, giving relevant ionic equations, the reactions you would expect to take place at the electrodes during the electrolysis of dilute sulfuric acid using inert graphite electrodes, and explain any changes which would occur in the electrolyte.

8 Compare the electrolysis of dilute sodium chloride solution with the electrolysis of concentrated sodium chloride solution, both using inert graphite electrodes.

Quantitative electrolysis

During electrolysis, the movement of electrons through the external circuit from anode to cathode results in a **flow of electrical charge** since each electron possesses an extremely small electrical charge. The quantity of a substance produced at, or dissolved from, an electrode during electrolysis is directly proportional to the **quantity of electrical charge**, or **quantity of electricity**, which flows through the electrolytic cell.

The quantity of electrical charge (***Q***), is measured in **coulombs (C)**. The quantity of electrical charge flowing through an electrolytic cell during electrolysis is dependent on **two** factors:

- The **current** (***I***), which is the rate of flow of the electrical charge. Current is measured in **amperes** (known as **amps, A**).
- The length of **time** (***t***) that the current flows for. Time is measured in **seconds** (**s**).

The quantity of electrical charge can be calculated using the formula below.

quantity of electrical charge (C) = current (A) × time (s)
or $Q = I \times t$

One mole of electrons (6.0×10^{23} electrons), has a total charge of **96 500 C**. This value is known as the **Faraday constant.**

*The **Faraday constant** is the size of the electrical charge on one mole of electrons, i.e.* ***96 500 C mol^{-1}***.

The following equations show **that one mole** of electrons is required to discharge **one mole** of an ion with a single charge:

$$\underset{}{M^+} + \underset{\textbf{1 mol}}{e^-} \longrightarrow M \quad \text{or} \quad N^- \longrightarrow N + \underset{\textbf{1 mol}}{e^-}$$

Using the Faraday constant, it follows that **96 500 C** is the quantity of electrical charge required to discharge **one mole** of ions with a **single charge**. The Faraday constant can be used to calculate the masses of substances and volumes of gases formed during electrolysis.

Example

To determine the mass of magnesium produced at the cathode when a current of 7.5 A flows through molten magnesium chloride for 25 minutes and 44 seconds.

Determine the **quantity of electricity (*Q*)** that flows:

Current = **7.5 A**

Time in seconds = (25 × 60) + 44 s = **1544 s**

Quantity of electricity (C) = current (A) × time (s)

∴ quantity of electricity = 7.5 × 1544 C

= **11 580 C**

Write the equation for the reaction at the **cathode**:

$$Mg^{2+}(l) + 2e^- \longrightarrow Mg(l)$$

2 mol **1 mol**

From the equation:

2 mol of electrons are required to form **1 mol Mg**

∴ 2 × 96 500 C are required to form 1 mol Mg

i.e. **193 000 C** form 1 mol Mg

∴ 1 C forms $\frac{1}{193\ 000}$ mol Mg

and 11 580 C form $\frac{1}{193\ 000} \times 11\ 580$ mol Mg

= **0.06 mol Mg**

Mass of 1 mol Mg = 24 g

∴ mass of **0.06 mol Mg** = 0.06 × 24 g

= **1.44 g**

Mass of magnesium produced = **1.44 g**

Sample questions

1 Calculate the volume of oxygen produced at the anode at rtp if an electric current of 5.0 A is passed through dilute sodium chloride solution for 3 hours, 51 minutes and 36 seconds.

Quantity of electricity that flows:

Current = **5.0 A**

Time in seconds = (3 × 60 × 60) + (51 × 60) + 36 s = **13 896 s**

∴ quantity of electricity = 5.0 × 13 896 C

= **69 480 C**

Equation for the reaction at the **anode**:

$$4OH^-(aq) \longrightarrow 2H_2O(aq) + O_2(g) + 4e^-$$

1 mol **4 mol**

From the equation:

4 mol electrons are lost in forming **1 mol O_2**

∴ 4 × 96 500 C form 1 mol O_2

i.e. **386 000 C** form 1 mol O_2

∴ 69 480 C form $\frac{1}{386\ 000} \times 69\ 480$ mol O_2

= **0.18 mol O_2**

Volume of 1 mol O_2 at rtp = 24.0 dm^3

∴ volume of **0.18 mol O_2** = 0.18 × 24.0 dm^3

= **4.32 dm^3**

Volume of oxygen produced = **4.32 dm^3**

2 A solution of copper(II) sulfate is electrolysed using inert graphite electrodes. Determine how long a steady current of 2.0 A must flow to cause the mass of the cathode to increase by 16.0 g.

Number of moles of copper produced at the cathode:

Mass of 1 mol Cu = 64 g

$\therefore$ number of moles in **16.0 g Cu** = $\frac{16.0}{64}$ mol

= **0.25 mol**

Equation for the reaction at the **cathode**:

$Cu^{2+}(aq) + 2e^- \longrightarrow Cu(s)$

2 mol **1 mol**

From the equation:

2 mol electrons are required to form **1 mol Cu**

$\therefore$ 2 × 96 500 C form 1 mol Cu

i.e. **193 000 C** form 1 mol Cu

$\therefore$ 0.25 × 193 000 C form 0.25 mol Cu

= **48 250 C**

i.e quantity of electricity required = **48 250 C**

Time taken for 48 250 C to flow using a current of 2.0 A:

Quantity of electricity (C) = current (A) × time (s)

48 250 C = 2.0 × time

$\therefore$ time = $\frac{48\,250}{2.0}$ s

= 24 125 s

= 6 hours 42 minutes 5 seconds

Time taken = <u>**6 hours 42 minutes 5 seconds**</u>

Industrial applications of electrolysis

Electrolysis is used in many ways in industry.

Extraction of metals from their ores

Electrolysis of the **molten ore** is used to **extract** aluminium, and metals above aluminium in the electrochemical series, from their ores (see p. 159).

Purification of metals (electrorefining)

Electrolysis is used to convert an impure metal into the pure metal, a process also known as **electrorefining**.

- The **anode** is the **impure metal**.
- The **cathode** is a very thin strip of the **pure metal**.
- The electrolyte is an aqueous solution containing **ions of the metal** being purified.

Example

Purification of copper

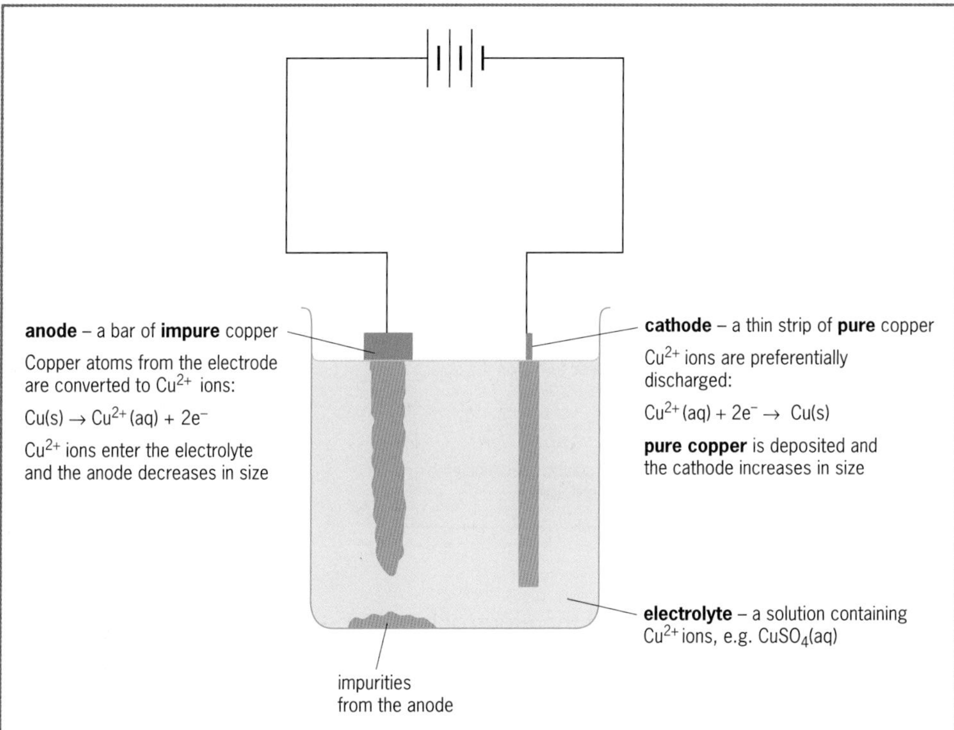

Figure 11.5 *The electrorefining of copper*

Note Electrorefining is only suitable to purify metals whose ions are **below hydrogen** in the electrochemical series.

Electroplating

Electroplating is the process by which a thin layer of one metal is deposited on another metal by electrolysis. It is used to **protect** the original metal from corrosion, to make it look more **attractive** or to make an inexpensive metal object appear more **valuable**.

Chrome plated rims

- The **anode** is a pure sample of the **metal** being used for plating.
- The **cathode** is the **object** to be electroplated.
- The electrolyte is an aqueous solution containing **ions of the metal** being used for plating.

Example

Silver plating

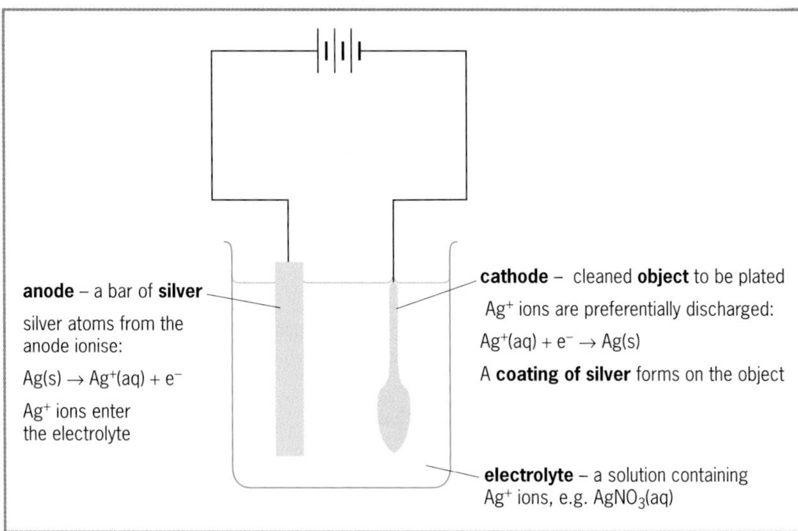

Figure 11.6 *Electroplating a spoon with silver*

Note Only metals whose ions are **below hydrogen** in the electrochemical series can be used for plating. **Silver**, **nickel** and **chromium** are the metals most commonly used.

Anodising

Anodising is a process used to increase the thickness of an unreactive oxide layer on the surface of a metal, usually the **aluminium oxide** (Al_2O_3) layer on the surface of aluminium objects. The aluminium oxide layer is relatively unreactive and adheres to the object, **protecting** it against corrosion. It also readily absorbs dyes, so can be attractively **coloured**.

- The **anode** is the cleaned **aluminium** object, such as a window frame or saucepan.
- The electrolyte is usually **dilute sulfuric acid**.

The aluminium anode **ionises** to form Al^{3+} ions:

$$Al(s) \longrightarrow Al^{3+}(aq) + 3e^-$$

At the same time, the SO_4^{2-} ions and the OH^- ions in the electrolyte move towards the anode. The Al^{3+} ions react with the OH^- ions and form a layer of aluminium oxide on the surface of the aluminium object:

$$2Al^{3+}(aq) + 3OH^-(aq) \longrightarrow Al_2O_3(s) + 3H^+(aq)$$

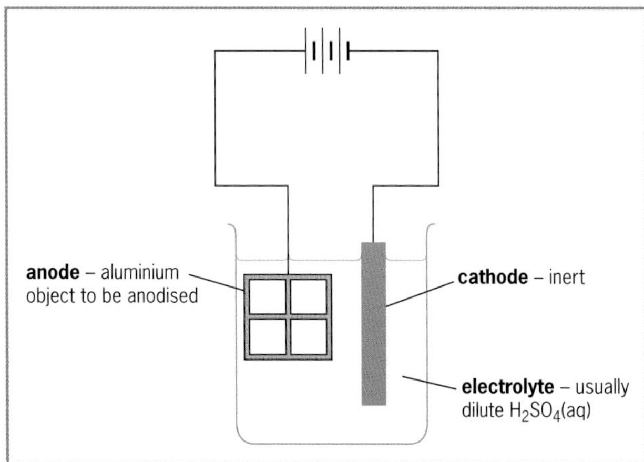

Figure 11.7 *Anodising an aluminium window frame*

Revision questions

9 What is the Faraday constant?

10 A current of 8.0 amperes flows for 48 minutes 15 seconds through a solution of copper(II) sulfate.

a What volume of oxygen, measured at rtp, would be produced at the anode?

b What mass of copper would be deposited at the cathode?

11 How long must a steady current of 0.5 amperes flow through the circuit in order to produce 112 cm^3 of chlorine at the anode during the electrolysis of concentrated sodium chloride solution at stp?

12 **a** Write the equations for the reactions occurring at the anode and cathode during the purification of a lump of silver.

b Calculate the mass of silver that could be purified if a current of 2.5 amperes flows for 4 hours 17 minutes 20 seconds during the purification process.

13 Explain, giving relevant equations, how electrolysis can be used to electroplate a spoon with nickel using nickel sulfate solution ($NiSO_4(aq)$) as the electrolyte.

14 What happens during the process of anodising?

12 Rates of reaction

The speed of chemical reactions varies considerably. Some reactions occur very **rapidly**, such as precipitation. Others occur very **slowly**, such as the rusting of iron and steel.

*The **rate of reaction** is a measured change in the concentration of a reactant or product with time, at a given temperature.*

The rate of a reaction can be determined by using the formulae given below.

$$\text{rate of reaction} = \frac{\text{decrease in the concentration of a reactant}}{\text{time taken for the decrease}}$$

or

$$\text{rate of reaction} = \frac{\text{increase in the concentration of a product}}{\text{time taken for the increase}}$$

Measuring rates of reaction

It is usually difficult to measure changes in concentration directly. Other **property changes** may occur which are more easily measured.

- The **volume of gas** produced over time can be measured using a gas syringe if the reaction produces a gas.
- The **decrease in mass** of the reaction can be measured over time using a balance if the reaction produces a gas that escapes.
- A change in **colour intensity**, **pressure**, **temperature** or **pH** can be measured over time if any of these properties change during a reaction.
- The appearance of a **precipitate** can be measured if one is formed in the reaction.

The collision theory for chemical reactions

During any chemical reaction, the existing bonds in the **reactants** must **break** so that new bonds can **form** in the **products**. In order to **react**:

- The particles of the reactants must **collide** with each other so that the bonds in the reactants can be broken.
- The reactant particles must collide with **enough energy** to break their bonds and enable new bonds to form in the products. This minimum energy is known as **activation energy** (see p. 112).
- The reactant particles must collide with the **correct orientation**. They must line up correctly with each other so that bonds can break and reform in the required way.

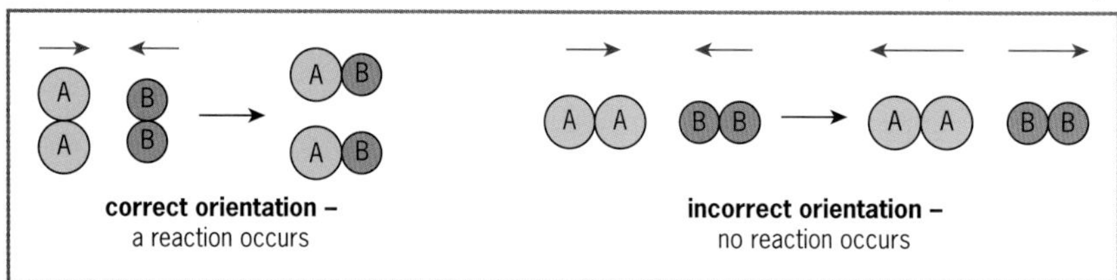

Figure 12.1 *Orientation of particles*

Not all collisions result in a reaction occurring. Some do not occur with the required activation energy and some do not occur with the correct orientation of particles. Any collision that results in a reaction is known as an **effective collision**.

Rate curves for reactions

A **rate curve** can be drawn if a measured property, such as concentration, is plotted on a graph against time as the reaction proceeds.

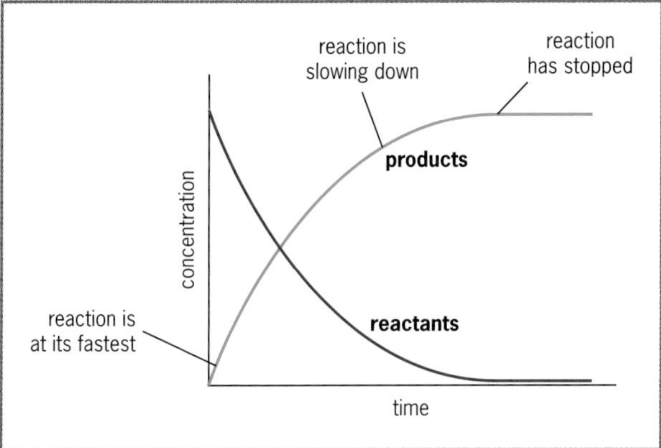

Figure 12.2 *Rate curves*

All rate curves have a very similar **shape**. This shape shows that the **rate** of a reaction **decreases** as the reaction proceeds.

- At the **beginning** of the reaction the gradient is at its **steepest**, showing that the rate is at its **highest**. The concentration of the reactant particles is at its highest resulting in the **frequency of collision** between particles being at its **highest**.
- As the **reaction proceeds** the gradient becomes **shallower**, showing that the rate is **decreasing**. The concentration of the reactant particles decreases as the reaction proceeds causing the **frequency of collision** between the particles to gradually **decrease**.
- The curve eventually becomes **horizontal** (its gradient becomes **zero**), showing the reaction has **reached completion** and **stopped**. At this point, one reactant has been used up and no more of its particles are left to collide. This reactant is known as the **limiting reactant** and its quantity determines the quantity of products made.

Factors that affect rates of reaction

The rate of a reaction is dependent on **four** main factors:

- **Concentration**.
- **Temperature**.
- **Surface area** (particle size).
- Presence or absence of a **catalyst**.

Pressure and **light** also affect the rate of some reactions. **Pressure** affects reactions if the reactants are in the **gaseous** state, such as the reaction between nitrogen and hydrogen to form ammonia. **Light** affects the reaction between methane and chlorine (see p. 137), hydrogen and chlorine, and photosynthesis in plants. In all these reactions, if pressure or light intensity **increase**, the rate of reaction **increases**.

Table 12.1 *Factors affecting the rates of reaction*

Factor	Effect on the rate of reaction	Explanation	Graphical representations and points to note
Concentration of reactants	The **higher** the concentration of a reactant, the **faster** the reaction. This applies to reactants in **solution**.	Increasing the concentration of a reactant increases the number of particles in a unit volume of solution. As a result, the particles **collide more frequently**, increasing the chances of effective collisions.	Rate of reaction against concentration: rate of reaction concentration 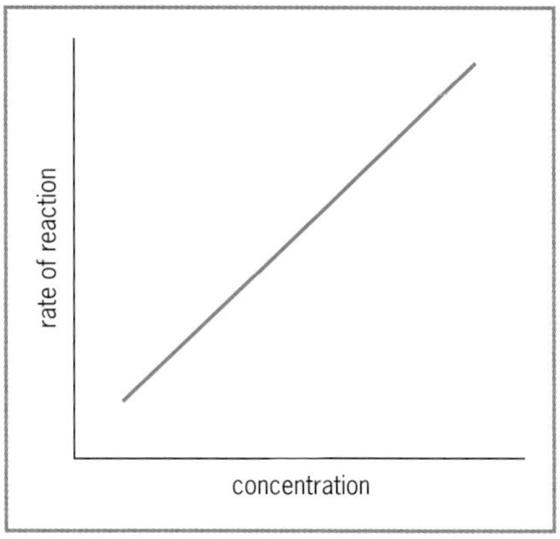
Temperature	The **higher** the temperature at which a reaction occurs, the **faster** the reaction. For some reactions, if the temperature increases by **10 °C**, the rate of the reaction approximately **doubles**.	Increasing the temperature of a reaction gives the reactant particles more **kinetic energy**. As a result: • The particles move faster so **collide more frequently**. • The particles collide with more energy so more collisions take place with **enough activation energy** for the particles to react. A combination of the two increases the chances of effective collisions.	Rate of reaction against temperature: rate of reaction temperature 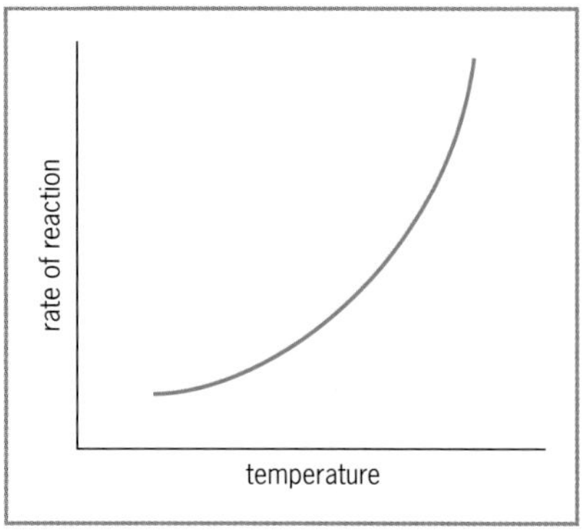
Surface area (particle size)	The **smaller** the particles of a reactant, the **faster** the reaction. This applies to reactants in the **solid** state.	The reaction occurs on the surface of a solid. Small solid particles have a larger total **surface area** than large particles with the same mass. Decreasing particle size exposes a greater surface area to the other reactant. As a result, the particles **collide more frequently**, increasing the chances of effective collisions.	In **flour mills** and **coal mines** the flour and coal dust are extremely flammable. A slight spark can start a reaction with the oxygen in the air which can be **explosive** because of the large surface area of the finely divided flour and coal particles.

Factor	Effect on the rate of reaction	Explanation	Graphical representations and points to note
Presence or absence of a catalyst	When added, most catalysts **speed up** a reaction. A few catalysts (inhibitors or negative catalysts) **slow down** a reaction.	A **catalyst** alters the rate of a reaction without itself undergoing any permanent chemical change. A catalyst speeds up a reaction by providing an alternative pathway that requires a **lower activation energy** than the normal pathway. As a result, more collisions occur with **enough activation energy** for the particles to react, increasing the chances of effective collisions.	**Enzymes** are biological catalysts. They speed up chemical reactions occurring in living cells. **Tetraethyl lead(IV) ($Pb(C_2H_5)_4$)** is an inhibitor that used to be added to petrol ('leaded petrol') to stop premature ignition ('knocking').

The effect of changing different factors on rate curves

The effect of changing any factor which alters the rate of a reaction can be shown on the **rate curve** for the reaction.

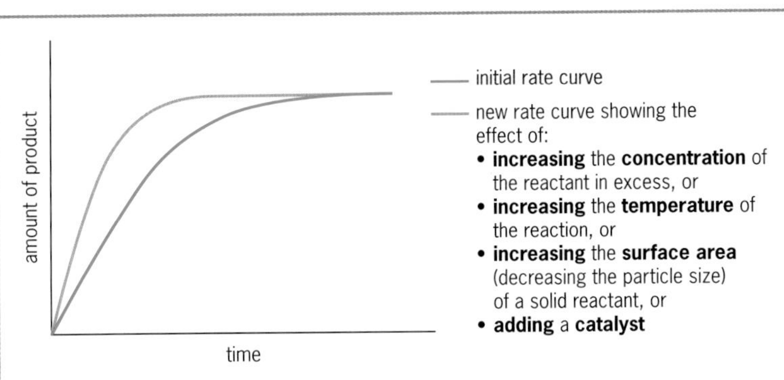

Figure 12.3 *The effect of changing any factor which increases the rate of a reaction*

Points to note from Figure 12.3

- The new curve has a **steeper gradient** indicating the reaction is occurring at a **faster rate**.
- The new curve becomes **horizontal sooner** indicating the reaction reaches completion in **less time**.
- Both curves become horizontal when the **same amount of product** has been made since the **number of moles** of the limiting reactant was unchanged.
- **Decreasing** the concentration of the reactant in excess, the temperature of the reaction, the surface area of a solid reactant, or adding a negative catalyst would cause the curve to have a **shallower gradient** than the original rate curve, become **horizontal later**, but become horizontal when the **same amount of product** has been made.

Revision questions

1 Define rate of reaction.

2 Explain the collision theory for chemical reactions.

3 **a** How does the rate of a reaction change as the reaction progresses?

b Explain why the rate of a reaction changes as it progresses.

4 Identify the FOUR main factors that can affect the rate of a reaction.

5 Explain how EACH of the following affects the rate of a reaction:

a increasing the concentration of a reactant

b increasing the temperature at which the reaction is carried out.

6 An experiment was carried out to determine how the rate of reaction between magnesium filings and hydrochloric acid varied with time. A known volume and concentration of hydrochloric acid was placed into a conical flask, the flask was placed on a balance and excess magnesium filings were added. The flask and its contents were weighed at regular intervals and their mass plotted against time, as shown in the graph below:

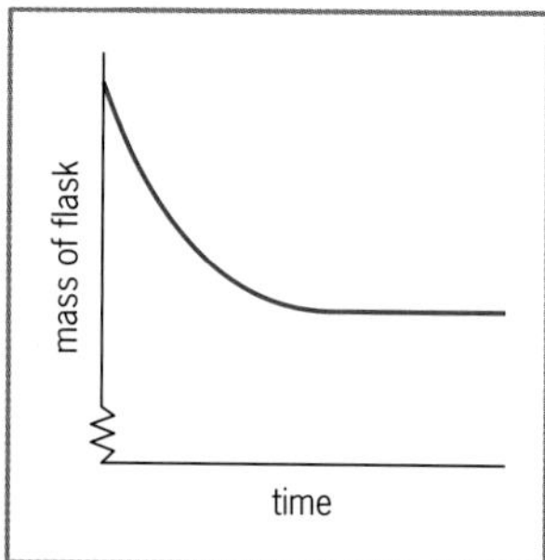

a On the graph above, draw a second curve to show the effect of repeating the experiment using magnesium ribbon.

b Explain the shape of the curve you drew in **a** above.

7 Why is it dangerous for a person to smoke a cigarette in a flour mill?

8 **a** What is a catalyst?

b How does a catalyst increase the rate of a reaction?

13 Energetics

All chemical substances contain energy stored in their bonds. When a chemical reaction occurs, there is usually a **change in energy** between the reactants and products. This is normally in the form of **heat energy**, but may also be in the form of light, nuclear or electrical energy.

Exothermic and endothermic reactions

Based on energy changes occurring, reactions can be of **two** types:

- An **exothermic reaction** produces heat which causes the reaction mixture and its surroundings to get **hotter** (it **releases energy** to the surroundings). Exothermic reactions include neutralisation reactions, burning fossil fuels and respiration in cells.
- An **endothermic reaction** absorbs heat which causes the reaction mixture and its surroundings to get **colder** (it **absorbs energy** from the surroundings). Endothermic reactions include dissolving certain salts in water, thermal decomposition reactions and photosynthesis in plants.

Breaking and forming bonds during reactions

During any chemical reaction, existing bonds in the **reactants** are **broken** and new bonds are **formed** in the **products**:

- Energy is **absorbed** when the existing bonds in the reactants are **broken**.
- Energy is **released** when new bonds are **formed** in the products:

reactants	⟶	**products**
existing bonds are **broken**		new bonds are **formed**
energy is **absorbed**		energy is **released**

- In an **exothermic reaction**:

  energy **absorbed** to break bonds < energy **released** when forming bonds

  The extra energy is **released** to the surroundings causing the **temperature** of the surroundings to **increase**.
- In an **endothermic reaction**:

  energy **absorbed** to break bonds > energy **released** when forming bonds

  The extra energy is **absorbed** from the surroundings causing the **temperature** of the surroundings to **decrease**.

Enthalpy changes during reactions

The energy content of a substance is called its **enthalpy** (***H***) and cannot be measured directly. However, it is possible to measure the **enthalpy change** (**Δ*H***) during a reaction.

The enthalpy change of a reaction is the difference between the enthalpy of the products and the enthalpy of the reactants:

enthalpy change of a **reaction** = (total enthalpy of **products**) – (total enthalpy of **reactants**)

or $$\Delta H_{reaction} = H_{products} - H_{reactants}$$

Δ*H* is usually expressed in kilojoules (**kJ**), or kilojoules per mol (**kJ mol^{-1}**).

- In an **exothermic reaction**:

  $$H_{products} < H_{reactants}$$

  The value of Δ*H* is less than zero. Δ*H* is **negative** (**–ve**). The extra energy from the reactants is **released** to the surroundings.

- In an **endothermic reaction**:

$$H_{products} > H_{reactants}$$

The value of ΔH is greater than zero. ΔH is **positive** (**+ve**). The extra energy gained by the products is **absorbed** from the surroundings.

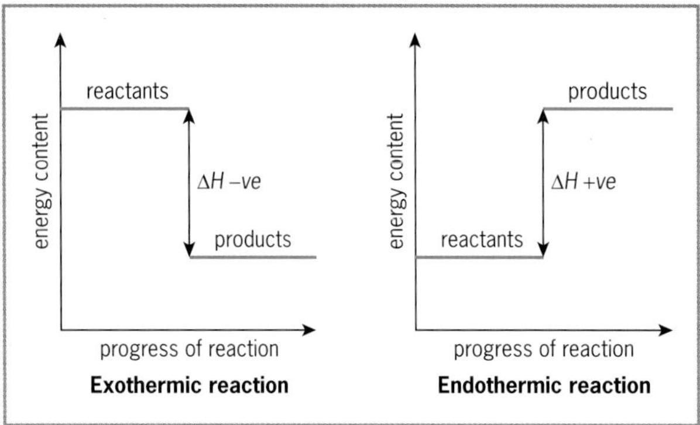

Figure 13.1 *Exothermic and endothermic reactions summarised*

Energy profile diagrams

An **energy profile diagram** can be drawn to illustrate the energy change during a chemical reaction. The diagram includes the **enthalpy** of the reactants and products, the **enthalpy change** (**ΔH**), and the **activation energy**. Activation energy can be thought of as the **energy barrier** of a reaction.

Activation energy *is the minimum amount of energy that reactants must be given, in excess of what they normally possess, so that bonds start breaking in the reactants and products start forming.*

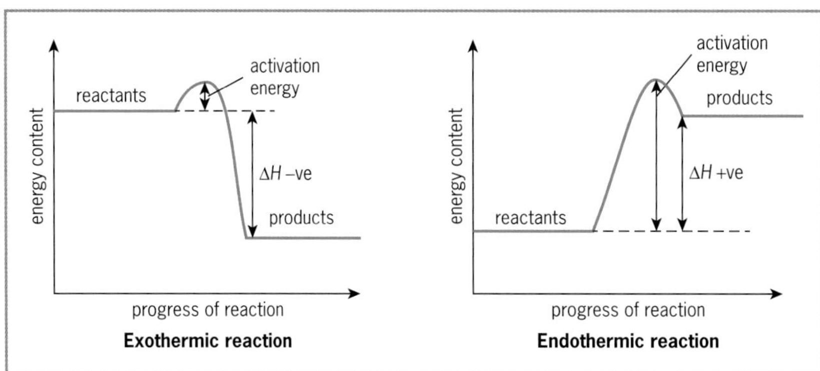

Figure 13.2 *Energy profile diagrams*

An energy profile diagram for a **specific reaction** must include:

- The **formulae** of the **reactants** and the **formulae** of the **products**.
- An arrow indicating **activation energy**.
- An arrow indicating **ΔH** with the **value of ΔH** written alongside.

Example

When **1 mol** of methane is completely burned in oxygen, 891 kJ of energy is **lost**, meaning that the reaction is **exothermic**:

$$CH_4(g) + 2O_2(g) \longrightarrow CO_2(g) + 2H_2O(g) \qquad \Delta H = -891 \text{ kJ mol}^{-1}$$

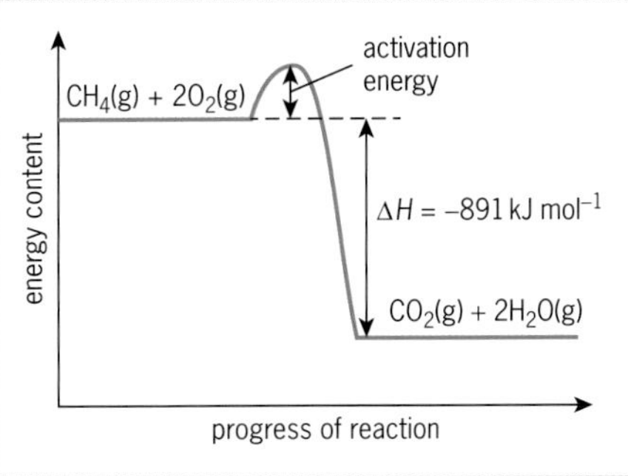

Figure 13.3 *Energy profile diagram for the combustion of methane*

The action of a catalyst

Most **catalysts** increase the rate of a reaction. A reaction in which a catalyst is used to increase the rate has a **lower activation energy** than the same reaction without a catalyst (see Table 12.1, p. 109). The effect of using a catalyst can be shown on energy profile diagrams.

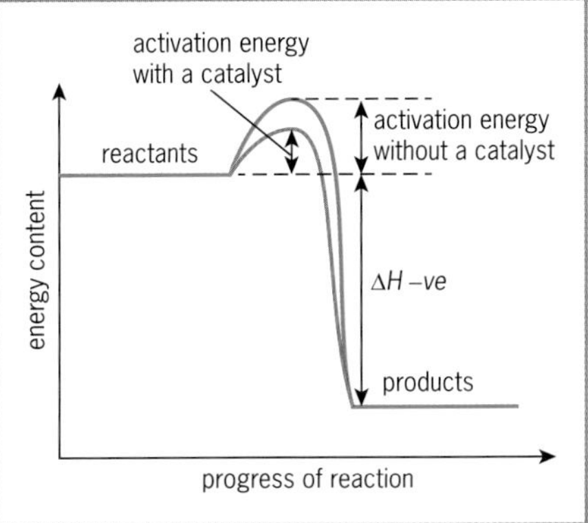

Figure 13.4 *Action of a catalyst on an exothermic reaction*

Calculating enthalpy changes

If the change in **temperature** that occurs during a reaction is measured, the **heat change**, known as the **heat of reaction**, can be determined using the formula given below.

heat change (ΔH)	= mass of reactants	× specific heat capacity	× temperature change
(J)	(g)	($J\ g^{-1}\ °C^{-1}$)	(°C)

Specific heat capacity is the quantity of heat energy required to raise the temperature of 1 g of a substance by 1 °C.

To determine the heat of reaction, the reaction is carried out in an insulated container called a **calorimeter**. The temperatures of the reactants are measured before mixing. The maximum or minimum temperature reached when the reactants are mixed is then measured and used to determine the **temperature change**. Three **assumptions** are made in calculating the heat of reaction:

- The **density** of a dilute aqueous solution is the same as pure water, **1 g cm^{-3}**. This means that 1 cm^3 of solution has a mass of 1 g.
- The **specific heat capacity** of a dilute aqueous solution is the same as pure water, **4.2 J g^{-1} °C^{-1}**. This means that it requires 4.2 J to increase the temperature of 1 g of water by 1 °C.
- A **negligible amount of heat** is lost to, or absorbed from, the surroundings during the reaction.

Determining the heat of solution

*The **heat of solution** is the heat change when 1 mol of solute dissolves in such a volume of solvent that further dilution by the solvent produces no further heat change.*

When a solute dissolves in a solvent:

- Bonds **break** between the solute particles; **ionic bonds** between ions break in ionic compounds and **intermolecular forces** between the molecules break in covalent substances. This **absorbs** energy from the surroundings.
- **Intermolecular forces** between the solvent molecules also **break**. This **absorbs** energy from the surroundings.
- Attractions form between the ions or molecules of the solute and the molecules of the solvent, a process is called **solvation.** This **releases** energy to the surroundings.

The reaction is **exothermic** if the energy **absorbed** to break bonds in the solute and solvent is **less** than the energy **released** during solvation.

The reaction is **endothermic** if the energy **absorbed** to break bonds in the solute and solvent is **greater** than the energy **released** during solvation.

When determining the heat of solution, the **initial** temperature of the water and the **maximum** or **minimum** temperature of the solution must be measured. The **temperature increase** or **decrease** and the **number of moles** of solute that dissolved must then be calculated.

Sample question

Dissolving 15.15 g of potassium nitrate in 100 cm³ of distilled water resulted in a temperature decrease of 10.2 °C. Calculate the heat of solution of potassium nitrate.

To determine the **number of moles** of KNO_3 dissolved:

Mass of 1 mol KNO_3 = 39 + 14 + (16 × 3) g = **101 g**

$\therefore$ number of moles in 15.15 g = $\frac{15.15}{101}$ mol

= **0.15 mol**

To determine the **heat of solution**:

Volume of water = 100 cm³

$\therefore$ mass of water = **100 g**

Final mass of solution = 100 + 15.15 g = **115.15 g**

Temperature decrease = **10.2 °C**

Specific heat capacity of the solution = **4.2 J g⁻¹ °C⁻¹**

Heat change = mass of solution × specific heat capacity × temperature change
(J) (g) (J g⁻¹ °C⁻¹) (°C)

$\therefore$ heat absorbed in dissolving **0.15 mol KNO_3** = 115.15 × 4.2 × 10.2 J

= **4933 J**

and heat absorbed in dissolving **1 mol KNO_3** = $\frac{4933}{0.15}$ J

= 32 887 J

= **32.9 kJ**

Heat of solution, ΔH = **+32.9 kJ mol⁻¹**

The heat of solution, ΔH, is **positive** because the temperature of the reaction **decreased** indicating that it **absorbed** energy from the surroundings. The reaction was **endothermic**.

Determining the heat of neutralisation

The ***heat of neutralisation*** *is the heat change when* ***1 mol of water*** *is produced in a neutralisation reaction between an alkali and an acid.*

When determining the heat of neutralisation, the temperature of both solutions must be measured and used to determine the **average initial** temperature. The **maximum** temperature of the solution after mixing must then be measured and used to calculate the **temperature increase**. Finally the **number of moles** of water made in the reaction must be determined.

Sample question

50 cm³ of sodium hydroxide solution with a temperature of 29.4 °C and concentration of 2.0 mol dm^{-3} is added to 50 cm³ of sulfuric acid of concentration 1.0 mol dm^{-3} and temperature 30.0 °C. The maximum temperature of the solution after mixing is 43.2 °C. Determine the heat of neutralisation.

To determine the **number of moles of water** made in the reaction:

1000 cm³ NaOH(aq) contains 2.0 mol NaOH

$\therefore$ 50 cm³ NaOH(aq) contains $\frac{2.0}{1000} \times 50$ mol NaOH

$= \mathbf{0.1\ mol\ NaOH}$

1000 cm³ H_2SO_4(aq) contains 1.0 mol H_2SO_4

$\therefore$ 50 cm³ H_2SO_4(aq) contains $\frac{1.0}{1000} \times 50$ mol H_2SO_4

$= \mathbf{0.05\ mol\ H_2SO_4}$

Equation: $2NaOH(aq) + H_2SO_4(aq) \longrightarrow Na_2SO_4(aq) + 2H_2O(l)$

i.e. 2 mol NaOH reacts with 1 mol H_2SO_4 forming 2 mol H_2O(aq)

$\therefore$ **0.1 mol NaOH** reacts with **0.05 mol H_2SO_4** to form **0.1 mol H_2O**

0.1 mol H_2O is made in the reaction

To determine the **heat of neutralisation**:

Total volume of solution = 50 + 50 = 100 cm³

$\therefore$ mass of solution = **100 g**

Average initial temperature $= \frac{29.4 + 30.0}{2}$ °C = 29.7 °C

Final temperature = 43.2 °C

$\therefore$ temperature increase = 43.2 – 29.7 °C = **13.5 °C**

Specific heat capacity of the solution = **4.2 J g^{-1} °C^{-1}**

Heat change = mass of solution × specific heat capacity × temperature change

(J) (g) (J g^{-1} °C^{-1}) (°C)

$\therefore$ heat evolved in forming **0.1 mol H_2O** = 100 × 4.2 × 13.5 J

= **5670 J**

and heat evolved in forming **1 mol H_2O** = $\frac{5670}{0.1}$ J

= 56 700 J

= **56.7 kJ**

Heat of neutralisation, ΔH = **–56.7 kJ mol^{-1}**

The heat of neutralisation, ΔH, is **negative** because the temperature of the reaction **increased**, indicating that it **lost energy** to the surroundings. The reaction was **exothermic**.

When any strong alkali reacts with any strong acid, the heat of neutralisation is always approximately **–57 kJ mol⁻¹**. This is because the energy change is for the common reaction occurring between the **OH^- ions** of the alkali and the **H^+ ions** of the acid:

$$OH^-(aq) + H^+(aq) \longrightarrow H_2O(l) \qquad \Delta H = -57\ kJ\ mol^{-1}$$

Revision questions

1 Distinguish between an exothermic reaction and an endothermic reaction.

2 Explain what happens during an exothermic reaction in terms of bonds breaking and bonds forming.

3 Hydrogen peroxide (H_2O_2) decomposes according to the following equation:

$$2H_2O_2(aq) \longrightarrow 2H_2O(l) + O_2(g) \qquad \Delta H = -98.2\ kJ\ mol^{-1}$$

The reaction can be speeded up by adding a catalyst.

a State, with a reason, if the reaction is exothermic or endothermic.

b Explain the change in enthalpy which occurs during the reaction.

c Draw a fully labelled energy profile diagram for the reaction.

d On your energy profile diagram, show the effect of adding a catalyst to the reaction.

4 What is the name of the main piece of apparatus used when determining heats of reaction?

5 Give the formula used to calculate heat change occurring during a reaction.

6 What are THREE assumptions that are made when calculating heats of reaction?

7 To determine the heat of solution of ammonium chloride, a student dissolved 7.49 g of the solid in 100 cm^3 of distilled water at 29.7 °C. He stirred the solution and recorded a minimum temperature of 22.8 °C.

a Define 'heat of solution'.

b State, with a reason, if the reaction was exothermic or endothermic.

c Calculate the heat of solution of ammonium chloride, assuming that the specific heat capacity of the solution is 4.2 $J\ g^{-1}\ °C^{-1}$.

d Draw a fully labelled energy profile diagram for the reaction.

8 **a** Define 'heat of neutralisation'.

b Explain why the heat of neutralisation for the reaction between sodium hydroxide and nitric acid has the same value as the heat of neutralisation for the reaction between potassium hydroxide and sulfuric acid.

Exam-style questions – Chapters 1 to 13

Structured questions

1 a) Figure 1 shows how the three states of matter can change from one form to another. Letters A, B, C and D represent the processes involved in these changes.

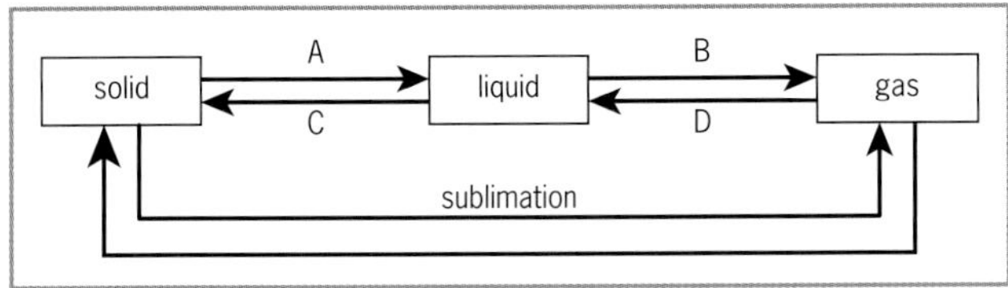

Figure 1 *Changes in state*

i) Identify the processes taking place at A, B, C and D. **(2 marks)**

ii) Describe what happens to the particles in a solid as it is heated and becomes a liquid. **(3 marks)**

iii) Identify ONE substance that sublimes. **(1 mark)**

b) A glass tube was set up as shown in Figure 2 and a white ring formed in the tube.

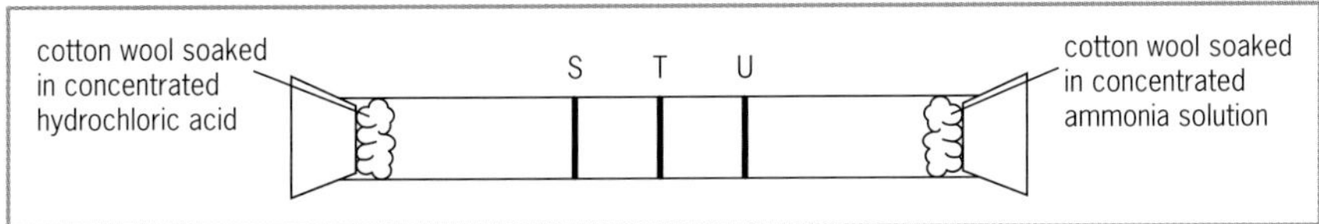

Figure 2 *Results of the experiment*

i) Give the name of the compound that made up the white ring. **(1 mark)**

ii) Write a chemical equation to represent the reaction that formed the compound named in **i)** above. **(1 mark)**

iii) At which point, S, T or U in Figure 2 did the white ring form? Give a reason for your answer. **(3 marks)**

c) By reference to particles, explain the reason for EACH of the following:

i) Oxygen gas readily takes the shape of the container it is in. **(2 marks)**

ii) You can tell when your sister is close by because you can smell her perfume. **(2 marks)**

Total 15 marks

2 a) Figure 3 shows the solubility curve for potassium dichromate(VI).

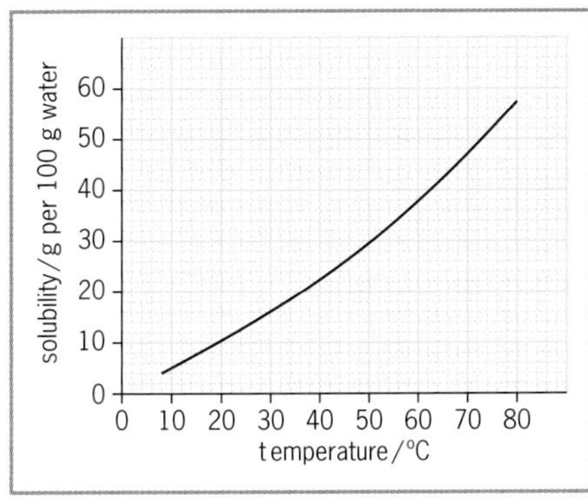

Figure 3 *The solubility curve for potassium dichromate(VI)*

i) Define the term 'solubility'. **(1 mark)**

ii) What can you deduce about the effect of temperature on the solubility of potassium dichromate(VI)? **(1 mark)**

iii) Use the graph in Figure 3 to determine the solubility of potassium dichromate(VI) at 44.0 °C. **(1 mark)**

iv) 250 g of distilled water is saturated with potassium dichromate(VI) at 62.0 °C. Determine the mass of potassium dichromate(VI) that would crystallise out of the solution if it is cooled to 26.0 °C. **(3 marks)**

b) Give TWO differences between a suspension and a colloid. **(2 marks)**

c) Whilst at the beach Jason collects a container full of sand and sea water and returns to the laboratory with the container.

i) Is the mixture that Jason collected homogeneous or heterogeneous? Give a reason for your answer. **(2 marks)**

ii) Plan and design an experiment that Jason could use to separate his mixture to obtain a dry sample of sand and pure water. Your answer should include the following:

– A suggested list of the apparatus he would use.

– An outline of the steps for the procedure that he would use. **(5 marks)**

Total 15 marks

3 Figure 4 shows part of the periodic table with helium (He), magnesium (Mg) and chlorine (Cl) in their correct positions. The positions of THREE unidentified elements are shown by the letters G, J and Q which are not the real symbols of the elements. Use this table to answer the following questions. When answering, use the letters given in the table as symbols of the unidentified elements; you are not expected to identify the elements.

Period	Group						
	I	II	III	IV	V	VI	VII	0
1								He
2								
3		Mg			G		Cl	
4		J					Q	
5								

Figure 4 *Part of the periodic table*

a) Give the electronic configuration of G. **(1 mark)**

b) i) State, with a suitable reason, whether J would react more or less vigorously with dilute hydrochloric acid than magnesium. **(3 marks)**

ii) Write a balanced equation for the reaction between magnesium and hydrochloric acid. **(2 marks)**

c) Compare chlorine and Q in terms of strength of oxidising power and provide a suitable explanation for your answer. **(3 marks)**

d) i) Identify TWO elements that would bond to form an ionic compound. **(1 mark)**

ii) Give the formula of the compound formed from the elements identified in **i)** above. **(1 mark)**

iii) Use a dot and cross diagram to illustrate the bonding between the elements identified in **i)** above. **(2 marks)**

e) Suggest TWO differences in the physical properties of elements J and Q. **(2 marks)**

Total 15 marks

4 Table 1 gives information about three particles. ^{18}X and ^{19}X are both atoms, $^{27}Y^{3+}$ is an ion.

Table 1 *Information about three particles*

Particle	Atomic number	Number of			Group number	Period number
		protons	neutrons	electrons		
^{18}X	9					
^{19}X						
$^{27}Y^{3+}$		13				

a) Complete Table 1 above by filling in the blank spaces. **(5 marks)**

b) What is the relationship between ^{18}X and ^{19}X? **(1 mark)**

c) Element X can form a compound with carbon, atomic number 6.

- **i)** What type of compound would carbon and element X form? **(1 mark)**
- **ii)** Give the formula of this compound. **(1 mark)**
- **iii)** Draw a dot and cross diagram to illustrate the formation of the compound between carbon and element X. **(2 marks)**

d) Y can also form a compound with element X.

- **i)** What type of compound would Y and X form? **(1 mark)**
- **ii)** Determine the relative atomic mass of this compound if it is formed between Y and ^{19}X. **(2 marks)**
- **iii)** Suggest TWO differences between the physical properties of the compound formed between carbon and X, and the compound formed between Y and X. **(2 marks)**

Total 15 marks

5 A neutralisation reaction is a reaction between a base and an acid. The neutralisation point in the reaction between an aqueous alkali and an aqueous acid can be determined by performing a titration.

a) i) What is meant by the term 'neutralisation point'? **(2 marks)**

- **ii)** What TWO ways can be used to determine the neutralisation point of a titration? **(2 marks)**

b) In order to determine the concentration of a solution of sulfuric acid, Susan placed the acid in a burette and performed a series of titrations. Figure 5 shows the burette readings before and after each titration using 25.0 cm^3 of potassium hydroxide solution of concentration 13.44 g dm^{-3}.

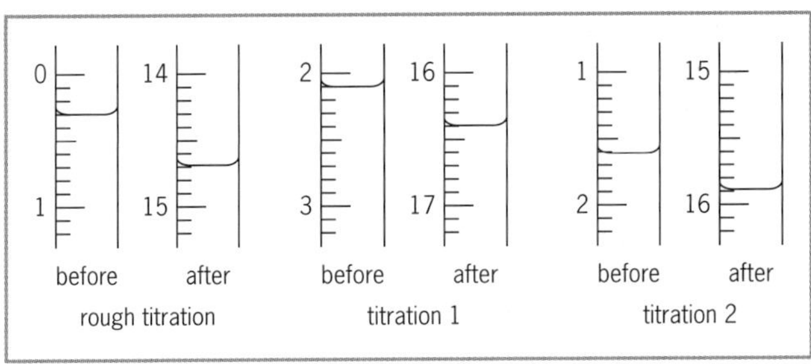

Figure 5 *Burette readings*

i) Complete the following table using the information in Figure 5.

	Titration number		
	Rough	1	2
Final burette reading/cm^3			
Initial burette reading/cm^3			
Volume of acid added/cm^3			

(3 marks)

ii) Determine the volume of sulfuric acid needed to neutralise 25.0 cm^3 of potassium hydroxide solution. **(1 mark)**

iii) Calculate the concentration of the potassium hydroxide solution in mol dm^{-3}. (Relative atomic masses: H = 1; O = 16; K = 39) **(1 mark)**

iv) Calculate the number of moles of potassium hydroxide used in the titration. **(1 mark)**

v) Write a balanced equation for the reaction. **(2 marks)**

vi) Determine the number of moles of sulfuric acid in the volume of solution used in the titration. **(1 mark)**

vii) Calculate the molar concentration of the sulfuric acid. **(1 mark)**

c) Identify ONE way we make use of neutralisation reactions in our daily lives. **(1 mark)**

Total 15 marks

6 The graph in Figure 6 shows the results of an experiment to determine the rate of a reaction between calcium carbonate crystals and nitric acid. The experiment was carried out at room temperature and pressure (rtp) by adding a fixed mass of crystals to excess acid of concentration 0.5 mol dm^{-3}.

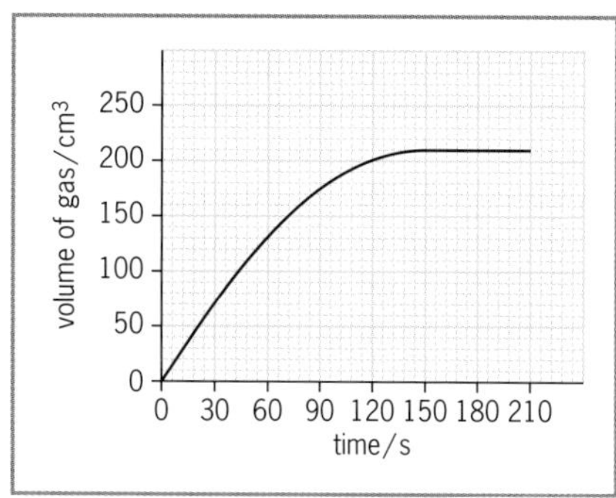

Figure 6 *Volume of gas produced against time*

a) i) Using the data from the graph determine the rate of the reaction in cm^3 s^{-1}:

– in the first minute

– in the second minute **(2 marks)**

ii) Explain why the rate of the reaction changes from the first minute to the second minute. **(4 marks)**

b) Write a balanced equation for the reaction. **(2 marks)**

c) Using information given in the graph:

i) Determine the volume of carbon dioxide evolved in the reaction. **(1 mark)**

ii) Determine the mass of calcium carbonate used. (Relative atomic masses: H = 1; C = 12; N = 14; O = 16; Ca = 40. Volume of 1 mol of gas rtp = 24 dm^3) **(2 marks)**

d) The experiment was repeated using the same mass of calcium carbonate crystals and excess nitric acid of concentration 0.75 mol dm^{-3}. Draw the expected results of this second experiment on the graph in Figure 6. **(2 marks)**

e) What effect, if any, would using the same mass of powdered calcium carbonate instead of crystals have on the rate of the evolution of carbon dioxide? Give a reason for your answer. **(2 marks)**

Total 15 marks

Extended-response questions

7 a) Table 2 gives some information about the physical properties of FOUR solid substances, EACH of which has a different structure.

Table 2 *The physical properties of four solid substances*

Substance	Melting point	Electrical conductivity	
		when solid	when molten
R	High	Good	Good
S	High	Poor	Good
T	Low	Poor	Poor
U	High	Poor	Poor

i) Suggest a possible structure for EACH substance. **(2 marks)**

ii) Account for the electrical conductivity of EACH of the substances, R and S. **(5 marks)**

iii) Suggest why U has a high melting point. **(2 marks)**

b) Diamond and graphite are both composed of carbon atoms. By referring to their bonding, explain why diamond is a non-conductor and is extremely hard whereas graphite is a good conductor of electricity and is soft and flaky. **(6 marks)**

Total 15 marks

8 a) Sulfuric acid can react with sodium hydroxide to form a normal salt and an acid salt.

i) Distinguish between a normal salt and an acid salt. **(2 marks)**

ii) Write ONE balanced equation EACH for the formation of a normal salt and an acid salt when sulfuric acid reacts with aqueous sodium hydroxide. **(3 marks)**

b) Epsom salt, composed of hydrated magnesium sulfate ($MgSO_4.7H_2O$), has many uses.

i) Give TWO uses of Epsom salt. **(2 marks)**

ii) What does the term 'hydrated' mean as applied to Epsom salt? **(1 mark)**

iii) Your teacher asks you to prepare a pure, dry sample of anhydrous magnesium sulfate in the laboratory starting with magnesium carbonate and 50 cm^3 of sulfuric acid of concentration 1.4 mol dm^{-3}. Outline the procedure you would use. **(3 marks)**

iv) Determine the maximum mass of anhydrous magnesium sulfate that you could make from your procedure.
(Relative atomic masses: H = 1; C = 12; O = 16; Mg = 24; S = 32) **(4 marks)**

Total 15 marks

9 a) Oxidation can be defined as an increase in oxidation number of an element due to the loss of electrons and reduction as a decrease in oxidation number of an element due to the gain of electrons. Explain how this statement applies in EACH of the following reactions.

i) $2I^-(aq) \longrightarrow I_2(aq) + 2e^-$ **(2 marks)**

ii) $Fe^{3+}(aq) + e^- \longrightarrow Fe^{2+}(aq)$ **(2 marks)**

b) 'Non-metals can behave as both oxidising and reducing agents'. Use the following reactions to support this statement.

i) $2ZnO(s) + C(s) \longrightarrow 2Zn(s) + CO_2(g)$ **(2 marks)**

ii) $Cl_2(g) + 2KBr(aq) \longrightarrow 2KCl(aq) + Br_2(aq)$ **(2 marks)**

c) A colourless solution, X, causes acidified potassium manganate(VII) solution to change from purple to colourless and aqueous potassium iodide to change from colourless to brown.

i) What can you deduce about solution X? **(1 mark)**

ii) Explain the colour change that occurred in EACH solution. **(4 marks)**

d) Granny buys some apples and cuts one into pieces but forgets to eat it. When she finally remembers, she finds that the cut surface of each piece has turned brown. Provide a suitable explanation for Granny's observation. **(2 marks)**

Total 15 marks

10 **a)** Using molten lead(II) bromide as an example of an electrolyte, describe what happens during the process of electrolysis when inert electrodes are used. Your answer should include relevant equations. **(4 marks)**

b) When copper(II) sulfate solution is electrolysed using graphite electrodes oxygen is produced at the anode; however, a different result is obtained when a copper anode is used. Explain the reason for this. **(4 marks)**

c) In some Caribbean countries steel pans are electroplated with chromium. Chromium(III) sulfate solution is usually used as the electrolyte.

i) Draw a labelled diagram to show how the apparatus would be set up to carry out the electroplating process. **(3 marks)**

ii) Determine the increase in mass of the steel pan if a current of 8.0 amperes flows for 8 hours 2 minutes 30 seconds during the electroplating process. (Relative atomic mass: Cr = 52; Faraday constant = 96 500 C mol^{-1}) **(4 marks)**

Total 15 marks

11 **a)** To determine the heat change which takes place when 1 mol of hydrochloric acid reacts completely with magnesium ribbon, Andrew added excess magnesium ribbon to 25 cm^3 of hydrochloric acid of concentration 0.5 mol dm^{-3} in a polystyrene cup. He recorded the following temperatures:

– initial temperature of the acid = 28.6 °C

– maximum temperature recorded on adding the magnesium ribbon = 49.8 °C

i) Assuming that the specific heat capacity of the solution is 4.2 J g^{-1} °C^{-1}, calculate the heat change for the reaction.
(Heat change = mass of acid × temperature change × specific heat capacity) **(3 marks)**

ii) State TWO other assumptions you have made in your calculation. **(2 marks)**

iii) Using your answer to **i)** above, calculate the heat change for reacting 1 mol of hydrochloric acid. **(2 marks)**

iv) Draw a fully labelled energy profile diagram for the reaction. **(3 marks)**

b) **i)** Write a balanced equation for EACH of the following reactions:

– aqueous sodium hydroxide reacting with sulfuric acid

– aqueous potassium hydroxide reacting with nitric acid **(3 marks)**

ii) Given that the heat of neutralisation is 57.2 kJ mol^{-1}, what would be the heat change for EACH of the above reactions? **(2 marks)**

Total 15 marks

14 Organic chemistry – an introduction

Organic chemistry is the study of compounds which contain **carbon**, known as **organic compounds**. Most organic compounds also contain hydrogen, many contain oxygen and some contain other elements such as nitrogen.

Bonding in organic compounds

A carbon atom has **four valence electrons** and can therefore form **four covalent bonds** with other carbon atoms, or atoms of other elements including hydrogen, oxygen, nitrogen and the halogens.

Example

Methane Formula: **CH_4**

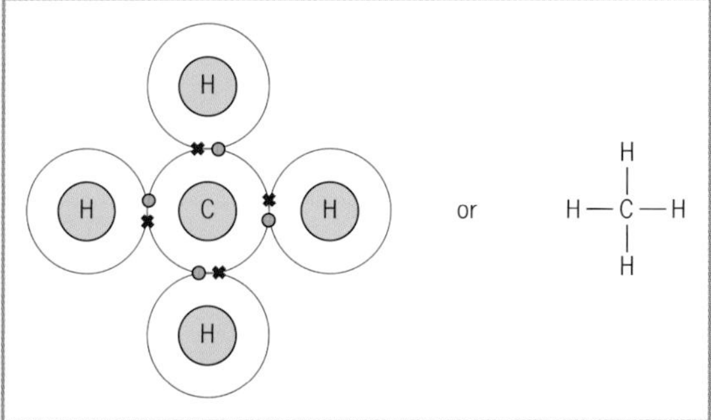

In the methane molecule, the four hydrogen atoms are arranged in a **tetrahedron** around the central carbon atom and the angle between covalent bonds is 109.5°:

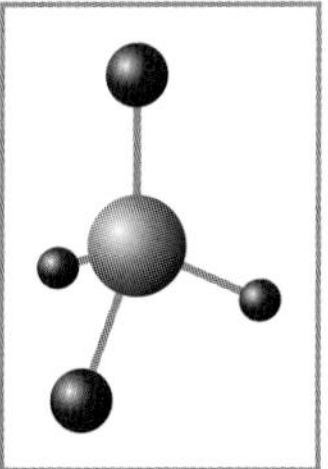

Figure 14.1 *Ball and stick model of methane*

Because carbon has four valence electrons, carbon atoms can bond with other carbon atoms in an almost unlimited way:

- **Single bonds** can form in which adjacent carbon atoms share **one pair** of electrons between them.

  e.g.
```
     H   H   H   H
     |   |   |   |
  H—C—C—C—C—H
     |   |   |   |
     H   H   H   H
```

  Any organic compound containing single bonds only between adjacent carbon atoms is referred to as being **saturated**.

- **Double bonds** can form in which adjacent carbon atoms share **two pairs** of electrons between them.

  e.g.
```
  H       H
   \     /
    C=C
   /     \
  H       H
```

  Any organic compound containing one or more double bonds between adjacent carbon atoms is referred to as being **unsaturated**.

- **Unbranched** and **branched chains** of carbon atoms of different lengths can form.

e.g.

```
    H  H  H  H              H  H  H  H  H
    |  |  |  |              |  |  |  |  |
 H—C—C—C—C—H            H—C—C—C—C—C—H
    |  |  |  |              |  |  |  |  |
    H  H  H  H              H  H  |  H  H
                                H—C—H
                                   |
                                   H
```

- **Rings** of carbon atoms can form with single or double bonds between adjacent carbon atoms.

e.g.

```
        H H                      H         H
     H  / \  H                    \       /
      \C—C/                        C=C
   H  /    \  H                   /     \
    \C      C/                H—C       C—H
   H/ \    / \H                   \\    //
      C—C                          C—C
    H/  / \  \H                   /     \
       H H                       H       H
```

Catenation is the ability of carbon atoms to bond covalently with other carbon atoms to form chains and rings of carbon atoms.

The structure of organic molecules

The simplest organic molecules can be thought of as being composed of **two** parts:

- The **hydrocarbon part** composed of only carbon and hydrogen atoms.
- The **functional group** (or **groups**), composed of a particular atom, group of atoms, or bond between adjacent carbon atoms, such as –OH, –COOH (see Table 14.2 on p. 126). The **chemical properties** of a compound are determined by the reactions of the functional group (or groups) present.

Formulae of organic compounds

The **formulae** of organic compounds can be written in different ways. The most common ways are:

- The **molecular formula**. This shows the **total number** of atoms of each element present in one molecule of the compound.
- The **fully displayed structural formula**. This shows how the atoms are bonded in one molecule in a **two-dimensional diagramatic** form, using a line to represent each covalent bond.
- The **condensed structural formula**. This shows the **sequence** and **arrangement** of the atoms in one molecule so that the position of attachment and nature of each functional group is shown without drawing the molecule.

  This can be condensed further to show the **total number** of **carbon** atoms and **total number** of **hydrogen** atoms in the hydrocarbon part of the molecule.

Example

Propanoic acid

The functional group of propanoic acid is –COOH.

- Its molecular formula is $C_3H_6O_2$
- Its fully displayed structural formula is

```
    H  H
    |  |   //O
 H—C—C—C
    |  |   \
    H  H    O—H
```

- Its condensed structural formula is CH_3CH_2COOH
  This can be condensed further to C_2H_5COOH

Homologous series

Organic compounds can be **classified** into groups known as **homologous series**. These are based on the functional group which they contain. Each homologous series can be represented by a **general formula**, for example, the general formula of the alkane series is C_nH_{2n+2}.

*A **homologous series** is a group of organic compounds which all possess the same functional group and can be represented by the same general formula.*

The characteristics of a homologous series

Each homologous series has the following characteristics:

- Members of a series all have the same **functional group**.
- Members of a series can all be represented by the same **general formula**.
- The **molecular formula** of each member of a series differs from the member directly before it or directly after it by $\mathbf{CH_2}$ and, therefore, by a relative molecular mass of **14**.
- Members of a series can all be **prepared** using the same general method.
- Members of a series all possess similar **chemical properties**. As the molar mass (number of carbon atoms per molecule) increases, the reactivity **decreases**.
- Members of a series show a gradual change in their **physical properties** as the number of carbon atoms per molecule increases. Generally, as molar mass increases, the melting point, boiling point and density **increase**.

How to name the straight chain members of a homologous series

Straight chain members of a homologous series have a name consisting of **two** parts:

- The first part, or **prefix**, which depends on the **total number of carbon atoms** in one molecule.

Table 14.1 *Prefixes used to name organic compounds*

Total number of carbon atoms	Prefix
1	meth-
2	eth-
3	prop-
4	but-
5	pent-
6	hex-

- The second part, which depends on the **functional group** present in the compound (see Table 14.2, p. 126).

Table 14.2 *The four main homologous series*

Name of homologous series	General formula	Functional group present	Naming members of the series	Example containing three carbon atoms	Fully displayed structural formula and condensed structural formula
Alkane	C_nH_{2n+2}	carbon–carbon single bond $-\overset{\mid}{\underset{\mid}{C}}-\overset{\mid}{\underset{\mid}{C}}-$	prefix + **ane**	**propane** (prop + ane)	H–C(H)(H)–C(H)(H)–C(H)(H)–H $CH_3CH_2CH_3$ or C_3H_8
Alkene	C_nH_{2n}	carbon–carbon double bond $>C=C<$	prefix + **ene**	**propene** (prop + ene)	H–C(H)(H)–C(H)=C(H)(H) $CH_3CH=CH_2$ or C_3H_6
Alcohol or alkanol	$C_nH_{2n+1}OH$	hydroxyl group $-O-H$ ($-OH$)	prefix + **anol**	**propanol** (prop + anol)	H–C(H)(H)–C(H)(H)–C(H)(H)–O–H $CH_3CH_2CH_2OH$ or C_3H_7OH
Alkanoic acid or carboxylic acid	$C_nH_{2n+1}COOH$	carboxyl group $-C(=O)-O-H$ ($-COOH$)	prefix + **anoic acid**	**propanoic acid** (prop + anoic acid)	H–C(H)(H)–C(H)(H)–C(=O)–O–H CH_3CH_2COOH or C_2H_5COOH

Structural isomerism

Organic compounds can have the same molecular formula but different structural formulae because their atoms are bonded differently. These are called **structural isomers**.

Structural isomers *are organic compounds which have the same molecular formula but different structural formulae.*

Structural isomerism *is the occurrence of two or more organic compounds with the same molecular formula but different structural formulae.*

Each different structural isomer has a **different name**, and if they contain the same functional group, they belong to the same homologous series.

Structural isomers of straight chain molecules can be formed in two ways:

- By the chain of carbon atoms becoming **branched**.
- By the **position** of the functional group changing.

Isomers formed by branching

Carbon chains can have **side branches** composed of one or more carbon atoms. For this to happen, the molecules must have four or more carbon atoms.

Example

Isomers of C_6H_{14}

C_6H_{14} has **five** isomers, three are given below:

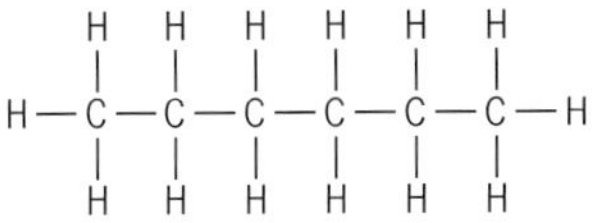

$CH_3CH_2CH_2CH_2CH_2CH_3$

A

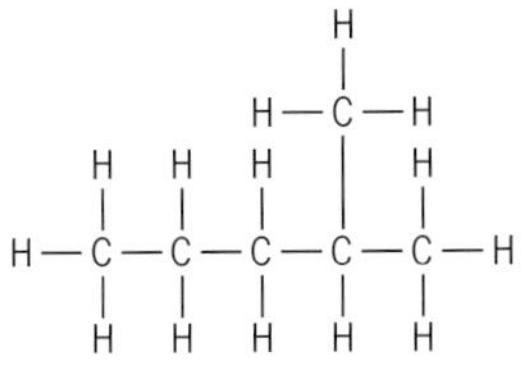

$CH_3CH_2CH_2CH(CH_3)CH_3$

B

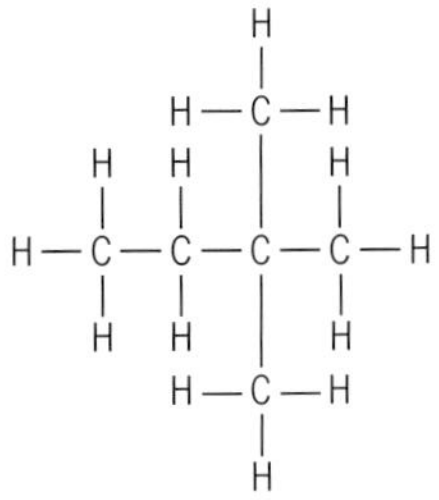

$CH_3CH_2C(CH_3)_2CH_3$

C

When drawing the structural formula of any organic compound, the longest continuous chain of carbon atoms must always be drawn **horizontally** and care must be taken not to draw bent versions of the straight chain isomer or mirror images of branched chain isomers.

To check if two structures are isomers, write the **condensed structural formula** of each. When reading these formulae forwards or backwards, if two formulae are the **same**, then the structures are **not** isomers.

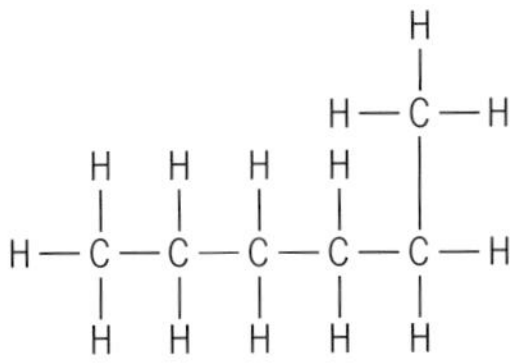

$CH_3CH_2CH_2CH_2CH_2CH_3$

This is a bent version of **A**, it is **not** another isomer

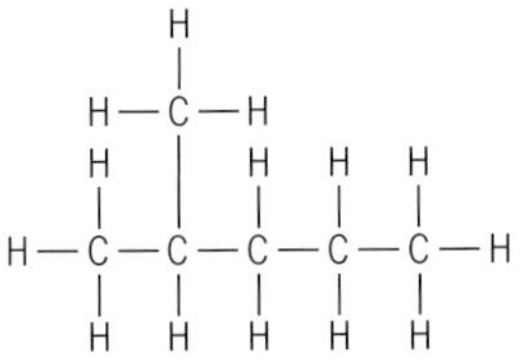

$CH_3CH(CH_3)CH_2CH_2CH_3$

This is a mirror image of **B**, it is **not** another isomer

How to name branched chain isomers

Side chains branching off from the longest chain of carbon atoms are called **alkyl groups**. They have the general formula C_nH_{2n+1} and are named using the appropriate prefix with the ending **'-yl'**.

Table 14.3 *Naming alkyl groups*

Formula of the alkyl group	Name
$-CH_3$	methyl
$-CH_2CH_3$ or $-C_2H_5$	ethyl
$-CH_2CH_2CH_3$ or $-C_3H_7$	propyl

The **name** of any branched chain molecule has **three** parts:

- The **first** part gives the **number of the carbon atom** to which the alkyl group (side chain) is attached.
- The **second** part gives the **name** of the **alkyl** group (see Table 14.3 above).
- The **third** part gives information about the **longest** continuous chain of carbon atoms. The **number of carbon atoms** in this chain is indicated using the correct **prefix**, and the **homologous series** to which the compound belongs is indicated using the correct **ending**.

Example

To determine the name of the following branched chain isomer of C_6H_{14}:

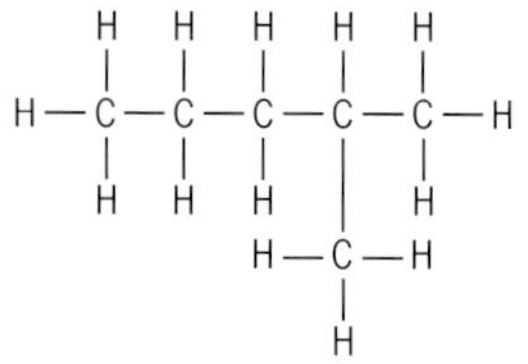

```
    H   H   H   H   H
    |   |   |   |   |
H — C — C — C — C — C — H
    |   |   |   |   |
    H   H   H   |   H
            H — C — H
                |
                H
```

Number the carbon atoms in the longest continuous chain of carbon atoms from the end closest to the alkyl group so that the group is attached to the atom with the **lowest** possible number:

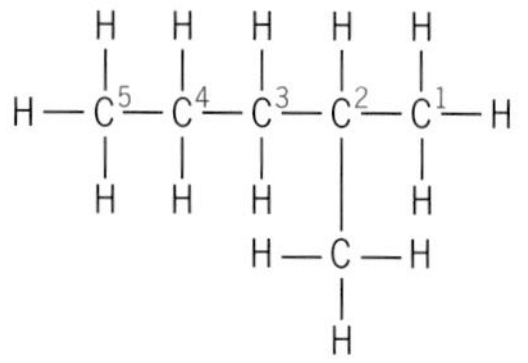

```
    H    H    H    H    H
    |    |    |    |    |
H — C5 — C4 — C3 — C2 — C1 — H
    |    |    |    |    |
    H    H    H    |    H
              H —  C — H
                   |
                   H
```

Name the isomer using the following information:

- The **first** part: the alkyl group is attached to carbon atom number **2**.
- The **second** part: the alkyl group has **one** carbon atom, so it is the **methyl** group.
- The **third** part: the longest continuous carbon chain has **five** carbon atoms, so its prefix is '**pent-**'. The compound has the general formula C_nH_{2n+2}, so it belongs to the **alkane** series which means its name ends in '**-ane**'.

The isomer is called **2-methylpentane**.

Isomers formed by changing the position of the functional group

A compound is usually drawn so that its functional group is shown at the right hand end of the molecule; however, its **position** can change. This is seen in the alkene and alcohol series (Tables 14.6, 14.7 and 14.8, pp. 130 and 131).

The alkanes: C_nH_{2n+2}

Alkanes contain only **single bonds** between carbon atoms. Alkanes with **four** or **more** carbon atoms show **structural isomerism** resulting from their ability to form **branched chains**.

Table 14.4 *The first three alkanes*

Molecular formula	CH_4	C_2H_6	C_3H_8
Structural formula and name	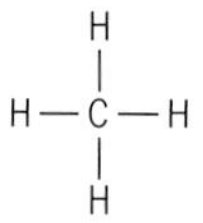 H | H—C—H | H **methane**	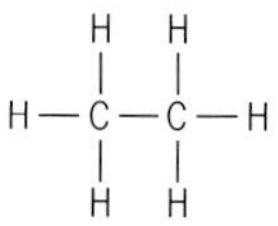 H H | | H—C—C—H | | H H **ethane**	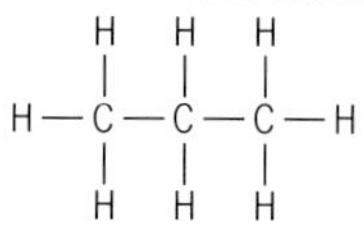 H H H | | | H—C—C—C—H | | | H H H **propane**

Table 14.5 *The isomers of alkanes with four, five and six carbon atoms*

Molecular formula	Structural formulae and names of isomers
C_4H_{10}	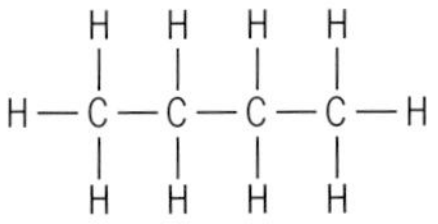 **butane** 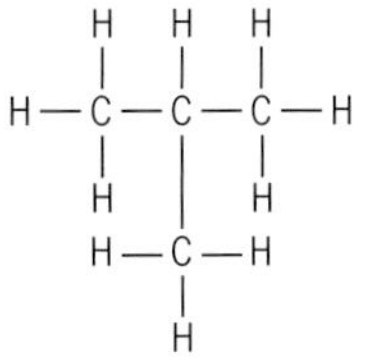 **2-methylpropane**
C_5H_{12}	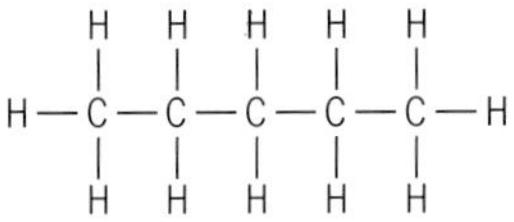 **pentane** 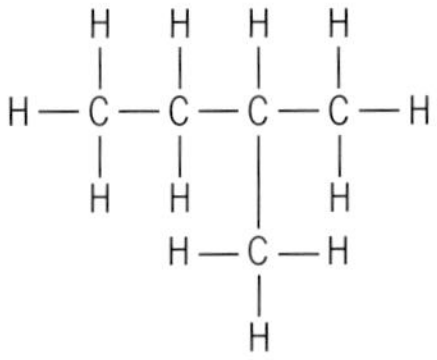 **2-methylbutane** 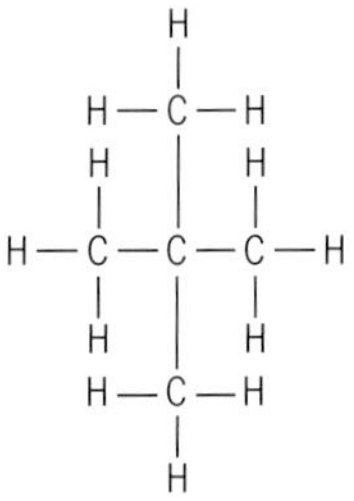 **2,2-dimethylpropane**
C_6H_{14}	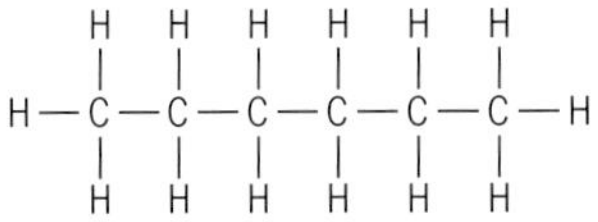 **hexane** 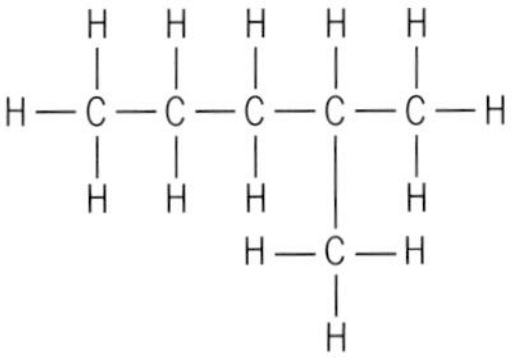 **2-methylpentane** 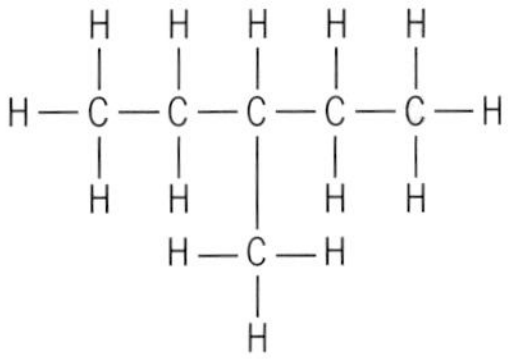 **3-methylpentane** 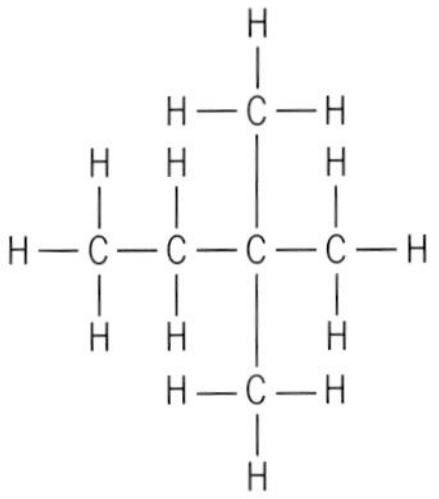 **2,2-dimethylbutane** 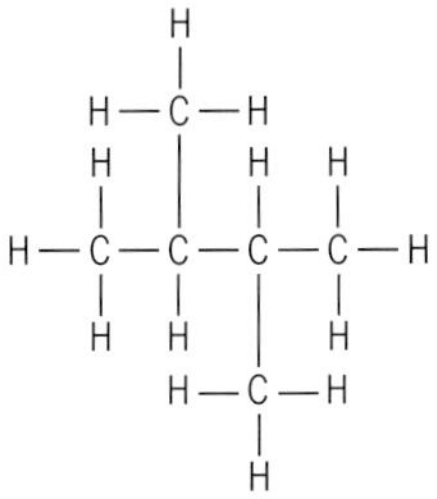 **2,3-dimethylbutane**

Note The prefix '**normal**' or '**n-**' is sometimes used before the name of the straight chain isomers. For example, the straight chain isomer of butane is sometimes called **normal butane** or **n-butane**.

The alkenes: C_nH_{2n}

Alkenes contain one double bond between two carbon atoms. Their functional group is this **carbon–carbon double bond**:

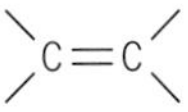

Alkenes with **four** or **more** carbon atoms show **structural isomerism** resulting from:

- A **change in position** of the carbon–carbon double bond. This bond must always be drawn horizontally.
- **Branching** of the molecule.

To name **unbranched isomers** of alkenes, **number** the carbon atoms in the longest continuous chain from the end closest to the double bond. Indicate the position of the double bond using the lowest possible number of the carbon atom it is attached to (see Tables 14.6 and 14.7). **Branched isomers** are named following the guidelines given on p. 127, but using the ending '**-ene**'.

Table 14.6 *The first three alkenes*

Molecular formula	Structural formulae and names		
C_2H_4	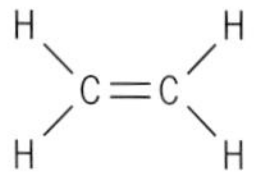 **ethene**		
C_3H_6	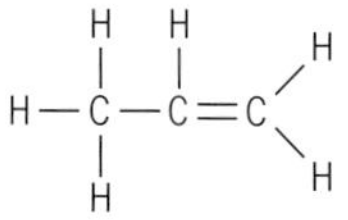 **propene**		
C_4H_8	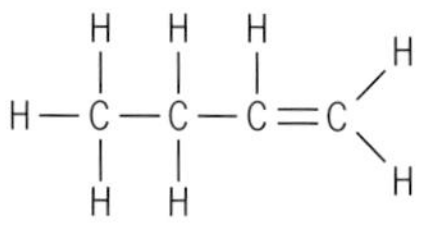 **but-1-ene** (or 1-butene) 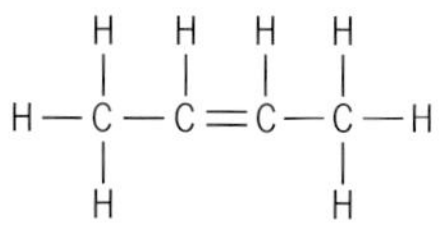	**but-2-ene** (or 2-butene) 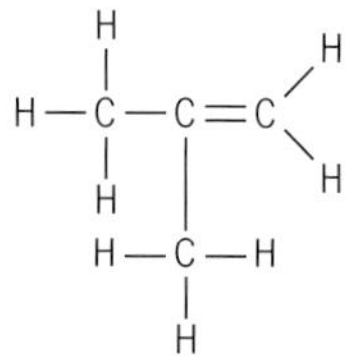	**2-methylpropene**

Table 14.7 *The unbranched isomers of pentene and hexene*

Molecular formula	Structural formulae and names		
C_5H_{10}	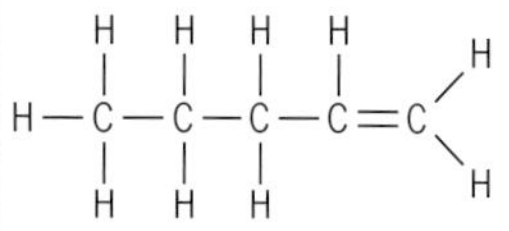 **pent-1-ene** (or 1-pentene) 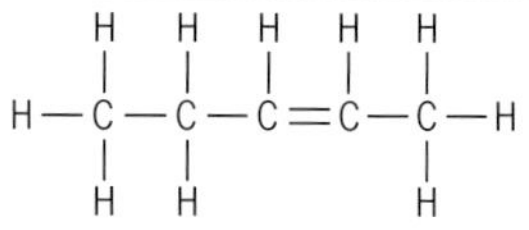	**pent-2-ene** (or 2-pentene)	
C_6H_{12}	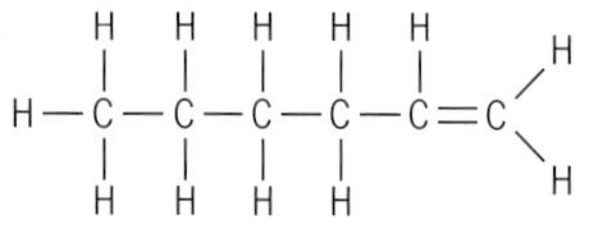 **hex-1-ene** (or 1-hexene) 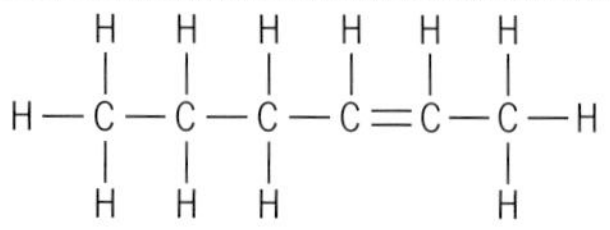	**hex-2-ene** (or 2-hexene) 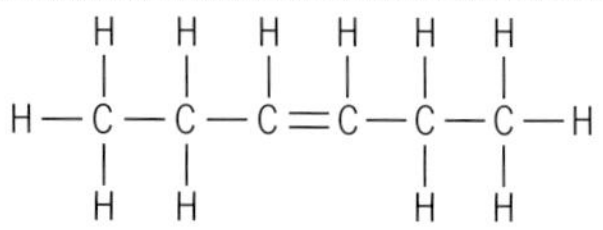	**hex-3-ene** (or 3-hexene)

Alcohols: $C_nH_{2n+1}OH$ or R–OH

Alcohols (or **alkanols**) have the **hydroxyl** group (–**OH**) as their functional group.

Alcohols with **three** or **more** carbon atoms show **structural isomerism** resulting from:

- A **change in position** of the hydroxyl (–OH) group.
- **Branching** of the molecule.

To name **unbranched isomers** of alcohols, **number** the carbon atoms in the longest continuous chain from the end closest to the –OH group. Indicate the position of the group using the number of the carbon atom it is bonded to (see Table 14.8). **Branched isomers** are named following the guidelines given on p. 127, but using the ending '**-anol**'.

Table 14.8 *The unbranched isomers of the first five alcohols*

<table>
<tr><th>Condensed formula</th><th colspan="3">Structural formulae and names</th></tr>
<tr><td>CH_3OH</td><td>H
|
H—C—O—H
|
H
methanol</td><td></td><td></td></tr>
<tr><td>C_2H_5OH</td><td>H H
| |
H—C—C—O—H
| |
H H
ethanol</td><td></td><td></td></tr>
<tr><td>C_3H_7OH</td><td>H H H
| | |
H—C—C—C—O—H
| | |
H H H
propan-1-ol
(1-propanol)</td><td>H H H
| | |
H—C—C—C—H
| | |
H OH H
propan-2-ol
(2-propanol)</td><td></td></tr>
<tr><td>C_4H_9OH</td><td>H H H H
| | | |
H—C—C—C—C—O—H
| | | |
H H H H
butan-1-ol
(or 1-butanol)</td><td>H H H H
| | | |
H—C—C—C—C—H
| | | |
H H OH H
butan-2-ol
(or 2-butanol)</td><td></td></tr>
<tr><td>$C_5H_{11}OH$</td><td>H H H H H
| | | | |
H—C—C—C—C—C—O—H
| | | | |
H H H H H
pentan-1-ol
(or 1-pentanol)</td><td>H H H H H
| | | | |
H—C—C—C—C—C—H
| | | | |
H H H OH H
pentan-2-ol
(or 2-pentanol)</td><td>H H H H H
| | | | |
H—C—C—C—C—C—H
| | | | |
H H OH H H
pentan-3-ol
(or 3-pentanol)</td></tr>
</table>

Note The unbranched isomers of **hexanol ($C_6H_{13}OH$)** are drawn and named in the same way as those of pentanol.

Alkanoic acids: $C_nH_{2n+1}COOH$ or R–COOH

Alkanoic acids (or **carboxylic acids**) have the **carboxyl** group (–**COOH**) as their functional group:

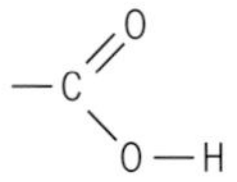

Table 14.9 *Names and formulae of the first six alkanoic acids*

Condensed formula	Structural formulae and names
HCOOH	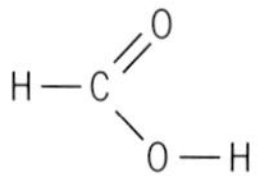 **methanoic acid**
CH_3COOH	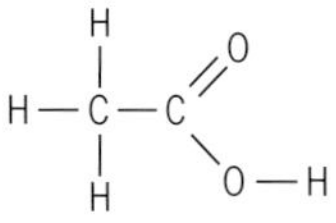 **ethanoic acid**
C_2H_5COOH	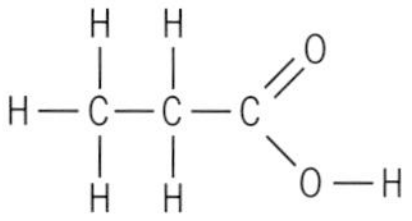 **propanoic acid**
C_3H_7COOH	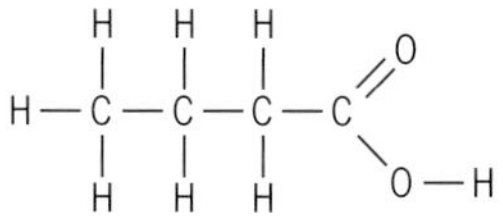 **butanoic acid**
C_4H_9COOH	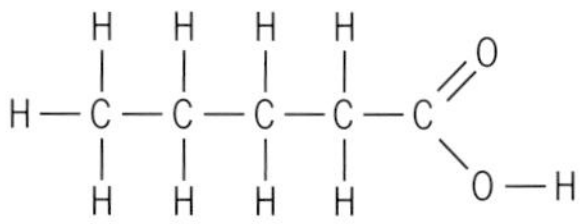 **pentanoic acid**
$C_5H_{11}COOH$	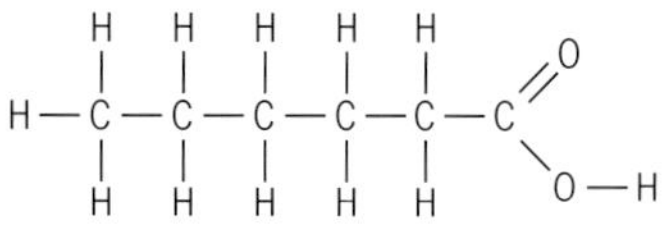 **hexanoic acid**

Revision questions

1 What are organic compounds?

2 **a** Why is carbon capable of forming a huge number of different organic compounds?

b Show, by means of a dot and cross diagram, how the atoms are bonded in the ethane molecule.

3 What is EACH of the following:

a a functional group **b** a fully displayed structural formula

c a homologous series?

4 **a** Give FOUR characteristics of a homologous series.

b Explain how straight chain members of a homologous series are named.

5 Give the general formula for members of EACH of the following homologous series:

a alcohol series **b** alkene series **c** alkanoic acid series **d** alkane series

6 Give the name and fully displayed structural formula of EACH of the following straight chain molecules:

a an alkene with THREE carbon atoms **b** an alcohol with ONE carbon atom

c an alkane with FOUR carbon atoms **d** an alkanoic acid with TWO carbon atoms

7 Name the homologous series to which EACH of the following compounds belongs and name EACH compound:

a $C_5H_{11}OH$ **b** C_2H_6 **c** C_2H_5COOH **d** C_4H_8

8 Define 'structural isomerism'.

9 Name the following compounds:

```
     H   H   H   H            H   H   H   H            H   H   H
     |   |   |   |            |   |   |   |   H        |   |   |
 H — C — C — C — C — H    H — C — C — C — C = C    H — C — C — C — H
     |   |   |   |            |   |   |       H        |   |   |
     H   |   H   H            H   H   H                H   OH  H
     H — C — H
         |
         H
         A                        B                        C
```

10 Draw the structural formula of EACH of the following:

a hex-3-ene **b** 2-methylpropene **c** 2,3-dimethylbutane

15 Sources of hydrocarbon compounds

Hydrocarbons are organic compounds composed of **carbon** and **hydrogen** atoms only. Alkanes and alkenes are both hydrocarbons.

Natural sources of hydrocarbons

There are **two** natural sources of hydrocarbons:

- **Natural gas**

  Natural gas is a mixture of the first four **alkanes**, mainly methane (CH_4) with small amounts of ethane (C_2H_6), propane (C_3H_8) and butane (C_4H_{10}). Before natural gas is sold commercially, the propane and butane are removed, leaving a mixture of **methane** and **ethane**. The **propane** and **butane** are then liquefied under pressure to produce liquefied petroleum gas, also known as LPG or 'bottled gas'.

- **Crude oil** (or **Petroleum**)

  Crude oil is a yellow to black oily liquid found in the earth. It is a complex mixture consisting of a large number of different solid and gaseous hydrocarbons dissolved in liquid hydrocarbons, mainly alkanes and some ringed compounds. To make it useful, petroleum is separated into its different components (or 'fractions') by **fractional distillation** at an oil refinery.

Fractional distillation of crude oil

Any impurities are first removed from the crude oil and then it is heated to about 400 °C. This produces a mixture of liquid and vapour which is piped into the bottom part of a **fractionating tower**. The **vapours** rise up the tower and the viscous liquid fraction, known as **bitumen** or **asphalt**, sinks to the bottom of the tower and is tapped off.

The vapours rising up the tower pass through a series of bubble caps and trays where they may **condense**. The **temperature decreases** going up the tower and the lower the boiling point of the hydrocarbon, the further the vapour will rise before condensing. The liquids produced when the vapours condense are tapped off at the different levels. Gases that do not condense are removed at the top of the tower as **refinery gas**. Each fraction tapped off is a mixture of hydrocarbons with similar sized molecules and boiling points. The different fractions have different **uses**. They are mainly used as fuels and lubricants, and to manufacture a variety of petrochemicals (see Figure 15.1, p. 135).

Fractionating towers at an oil refinery

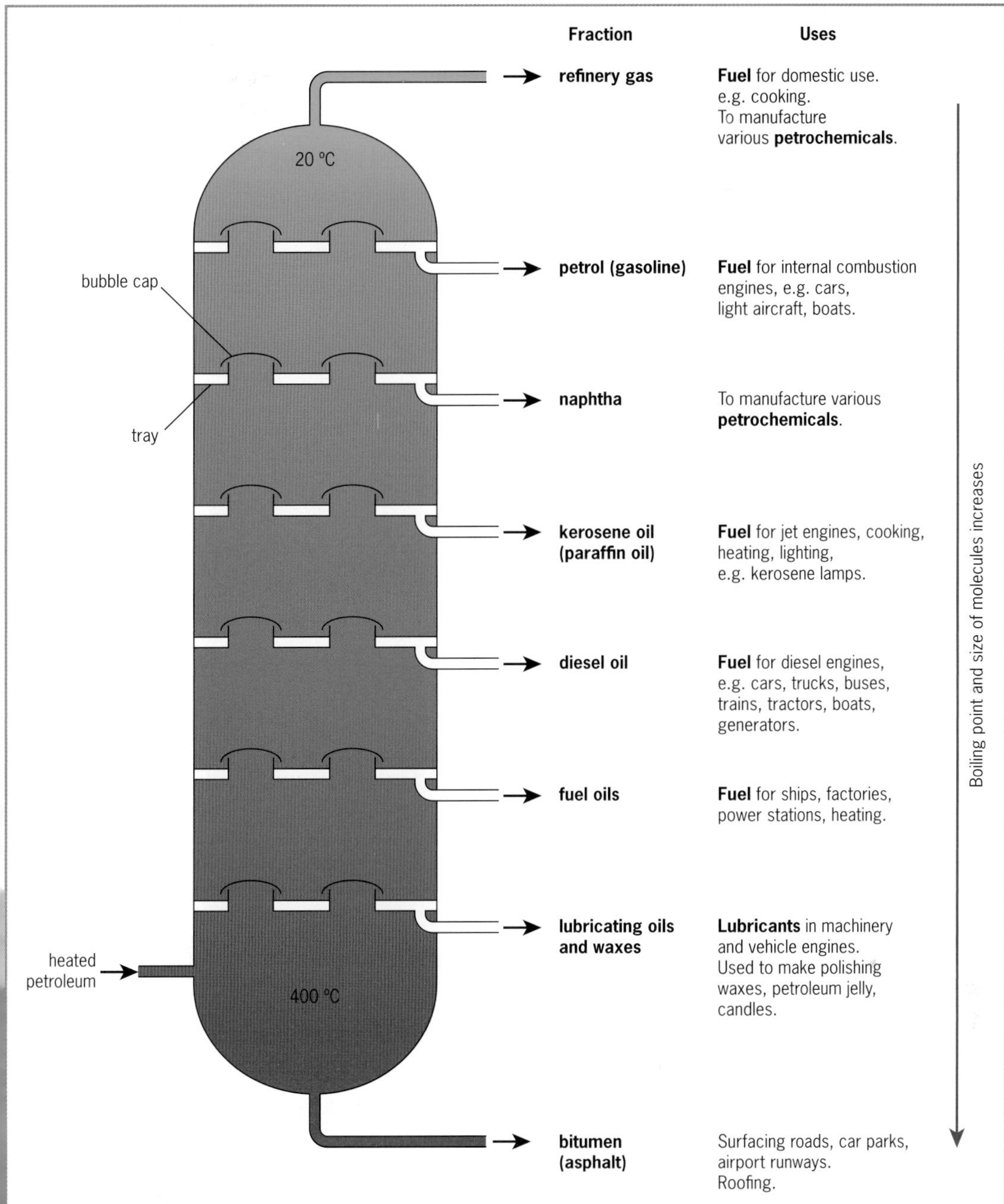

Figure 15.1 *Uses of the fractions produced by fractional distillation of crude oil*

Cracking hydrocarbons

Cracking *is the process by which long chain hydrocarbon molecules are broken down into shorter chain hydrocarbon molecules by breaking carbon–carbon bonds.*

Cracking is **important** for **two** reasons:

- Cracking **increases** the production of the **smaller**, more useful hydrocarbons, such as petrol. Fractional distillation produces an excess of the larger hydrocarbon molecules and insufficient of the smaller ones to meet current demands.

- Cracking produces the very reactive **alkenes** which are used in the **petrochemical industry** to make many other useful organic compounds. Fractional distillation does not produce alkenes, whereas cracking always results in the formation of at least **one alkene**.

Cracking can be carried out in **two** ways:

- **Thermal cracking**, which uses temperatures up to about 750 °C and pressures up to 70 atmospheres.
- **Catalytic cracking**, which uses temperatures of about 500 °C, at fairly low pressures in the presence of a **catalyst**.

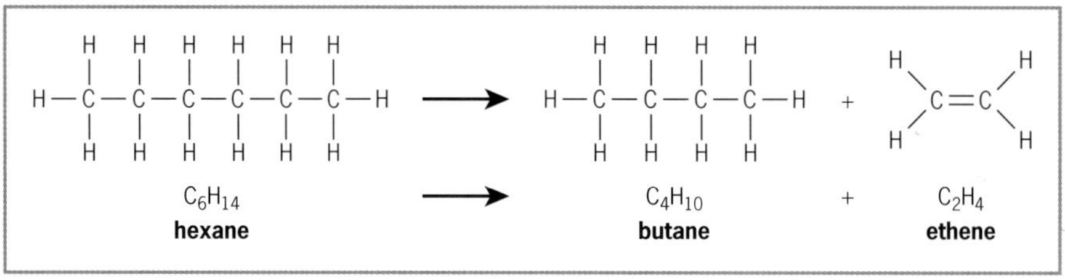

$$C_6H_{14} \longrightarrow C_4H_{10} + C_2H_4$$

hexane → **butane** + **ethene**

Figure 15.2 *One possible way of cracking hexane*

Revision questions

1. What are hydrocarbons?
2. Identify TWO natural sources of hydrocarbons.
3. Outline the process by which crude oil is separated into different fractions.
4. Name FOUR fractions obtained when crude oil is fractionally distilled, and give ONE major use of EACH fraction named.
5. **a** What happens when hydrocarbons are cracked?

   b Give TWO reasons why cracking hydrocarbons is important.
6. Pentane (C_5H_{12}) can be cracked in THREE different ways to produce TWO different compounds in EACH case. Show, by means of chemical equations, the THREE different ways pentane can be cracked and name the products in EACH case.

16 Reactions of carbon compounds

The **chemical reactions** of carbon compounds are determined by the reactions of the **functional group** (or groups) present in the compounds.

The alkanes: C_nH_{2n+2}

Alkanes are **saturated** hydrocarbons, meaning they have only **single bonds** between adjacent carbon atoms. Alkanes with 1 to 4 carbon atoms in their molecules are **gases** at room temperature, those with 5 to 16 carbon atoms are **liquids** and those with 17 or more carbon atoms are **solids**.

Alkanes are relatively **unreactive** because the carbon–carbon single bonds in their molecules are strong and not easily broken.

Reactions of alkanes

- **Alkanes burn easily in air or oxygen**

  Alkanes burn in air or oxygen to form **carbon dioxide** and **water** as steam. They burn with **clear**, **blue**, **non-smoky flames** because they have a **low** ratio of carbon to hydrogen atoms in their molecules. All the carbon is converted to carbon dioxide and no unreacted carbon remains in the flames to make them smoky. The reactions are **exothermic**, producing large amounts of heat energy.

  e.g. $CH_4(g) + 2O_2(g) \longrightarrow CO_2(g) + 2H_2O(g) \quad \Delta H\ -ve$

- **Alkanes undergo substitution reactions with halogens**

  Under the correct conditions, alkanes undergo **substitution reactions** with **halogens**. In these reactions, the hydrogen atoms in the alkane molecules are replaced by halogen atoms such as chlorine or bromine. For the reaction to occur, energy in the form of **light** is required; ultraviolet light works best. The products of the halogenation of alkanes are known as **haloalkanes** or **alkyl halides**.

Example

The reaction between methane and chlorine

In the dark, **no** reaction occurs. In bright light, a **rapid** reaction occurs. In dim light, a **slow substitution** reaction occurs in stages, where one hydrogen atom is replaced by one chlorine atom at a time:

$$CH_4(g) + Cl_2(g) \xrightarrow{\text{light}} CH_3Cl(g) + HCl(g)$$

monochloromethane; hydrogen chloride

$$CH_3Cl(g) + Cl_2(g) \xrightarrow{\text{light}} CH_2Cl_2(l) + HCl(g)$$

dichloromethane

$$CH_2Cl_2(l) + Cl_2(g) \xrightarrow{\text{light}} CHCl_3(l) + HCl(g)$$

trichloromethane

$$CHCl_3(l) + Cl_2(g) \xrightarrow{\text{light}} CCl_4(l) + HCl(g)$$

tetrachloromethane

The overall reaction:

$$CH_4(g) + 4Cl_2(g) \xrightarrow{\text{light}} CCl_4(l) + 4HCl(g)$$

Similar substitutions occur with **bromine vapour** or **bromine solution** and with other alkanes, though the reactions are slower. During the reaction between bromine and any alkane, the colour of the bromine **slowly** fades from **red brown** to **colourless** in the presence of **UV light**. This indicates that the bromine is being used up in the substitution reaction.

The alkenes: C_nH_{2n}

Alkenes are **unsaturated** hydrocarbons because they each contain one **carbon–carbon double bond**. The presence of this double bond as their functional group makes alkenes **more reactive** than alkanes.

Reactions of alkenes

- **Alkenes burn easily in air or oxygen**

  Alkenes burn in air or oxygen to form **carbon dioxide** and **water** as steam. They burn with **smoky yellow flames** because they have a higher ratio of carbon to hydrogen atoms in their molecules than alkanes. Not all the carbon is converted to carbon dioxide and the unreacted carbon remains, giving the flames a yellow, smoky appearance. The reactions are **exothermic**.

  e.g. $C_2H_4(g) + 3O_2(g) \longrightarrow 2CO_2(g) + 2H_2O(g) \quad \Delta H \text{ –ve}$

- **Alkenes undergo addition reactions**

  Alkenes undergo **addition reactions** with other small molecules in which the alkene and the other molecule react to form **one** molecule. One bond in the double bond is broken and the compound formed is **saturated** (it has no double bonds).

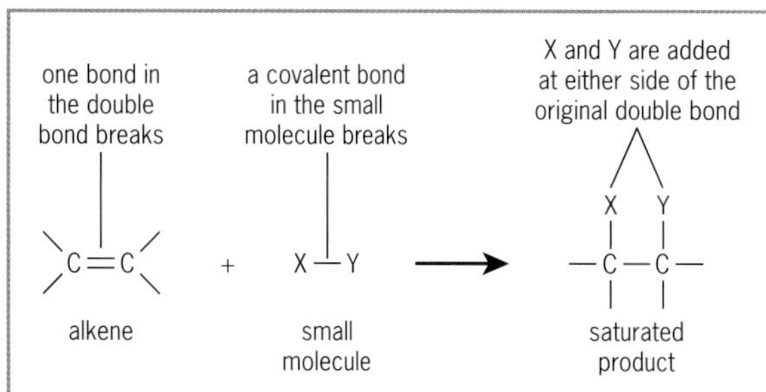

Figure 16.1 *Summary of addition reactions of alkenes*

Table 16.1 *Addition reactions of alkenes*

Reaction	Equation and conditions for reactions of ethene	Product
Addition of **hydrogen**	$H_2C{=}CH_2$ (ethene) + H—H (hydrogen) $\xrightarrow[\text{5 atm, 150 °C}]{\text{Ni catalyst}}$ $H_3C{-}CH_3$ (**ethane**) or $C_2H_4(g) + H_2(g) \xrightarrow[\text{5 atm, 150 °C}]{\text{Ni catalyst}} C_2H_6(g)$	alkanes

Reaction	Equation and conditions for reactions of ethene	Product
Addition of **halogens** (chlorine gas, bromine vapour, or a solution of bromine in water or tetrachloromethane)	$H_2C{=}CH_2 + Br{-}Br \longrightarrow H{-}CHBr{-}CHBr{-}H$ ethene + bromine → **1,2-dibromoethane** or $C_2H_4(g) + Br_2(g) \longrightarrow C_2H_4Br_2(l)$ The bromine vapour or bromine solution changes colour **rapidly** from **red-brown** to **colourless.**	haloalkanes
Addition of **hydrogen halides**, (hydrogen chloride, hydrogen bromide or hydrogen iodide)	$H_2C{=}CH_2 + H{-}I \longrightarrow H{-}CH_2{-}CHI{-}H$ ethene + hydrogen iodide → **monoiodooethane** or $C_2H_4(g) + HI(g) \longrightarrow C_2H_5I(l)$	haloalkanes
Addition of **water** in the form of steam	$H_2C{=}CH_2 + H{-}OH \xrightarrow[\text{70 atm, 300 °C}]{H_3PO_4\text{ in sand catalyst}} H{-}CH_2{-}CH_2{-}O{-}H$ ethene + water → **ethanol** or $C_2H_4(g) + H_2O(g) \xrightarrow[\text{70 atm, 300 °C}]{H_3PO_4\text{ in sand catalyst}} C_2H_5OH(l)$	alcohols
Addition reaction with **acidified potassium manganate(VII) solution**	$H_2C{=}CH_2 + H{-}OH + [O] \longrightarrow H{-}O{-}CH_2{-}CH_2{-}O{-}H$ ethene + [O] from oxidising agent → **ethane-1,2-diol** or $C_2H_4(g) + H_2O(l) + [O] \longrightarrow C_2H_4(OH)_2(l)$ The acidified potassium manganate(VII) solution changes colour rapidly from **purple** to **colourless**. The alkene **reduces** the purple MnO_4^- ion to the colourless Mn^{2+} ion.	dialcohols
Polymerisation	When exposed to high pressure, a catalyst and heat, alkene molecules undergo addition reactions with each other to form large molecules called **polyalkenes**, e.g. ethene forms **polyethene** (see p. 149).	polyalkenes

Distinguishing between an alkane and an alkene

An alkane can be distinguished from an alkene by reacting both with **bromine** solution or **acidified potassium manganate(VII)** solution. Both reactions test to see if the compound contains a carbon–carbon double bond causing it to be **unsaturated**.

Table 16.2 *Distinguishing between an alkane and an alkene*

Test	Observations and their explanations	
	Alkane	Alkene
Add **bromine solution**, i.e. bromine dissolved in water (bromine water) or in tetrachloromethane	The solution stays **red-brown.** **No reaction** occurs between the alkane and bromine under normal laboratory conditions.	The solution **rapidly** changes colour from red-brown to **colourless.** A **rapid addition reaction** occurs between the alkene and bromine solution forming a **colourless** haloalkane.
Add **acidified potassium manganate(VII) solution**	The solution stays **purple.** **No reaction** occurs between the alkane and acidified potassium manganate(VII) solution.	The solution **rapidly** changes colour from purple to **colourless.** A **rapid addition reaction** occurs during which the alkene reduces the purple MnO_4^- ion to the **colourless** Mn^{2+} ion.

Note In **UV light**, alkanes cause a solution of bromine to **slowly** change colour from red-brown to **colourless** due to the **slow substitution** reaction occurring.

Uses of alkanes and alkenes

Uses of alkanes

- Alkanes are used as **fuels** for the following reasons:
  - They **burn** very easily.
  - They release large amounts of **energy** when they burn because the reactions are exothermic.
  - They burn with **clean blue flames** which contain very little soot.
  - They are easy to **store**, **transport** and **distribute.**
- Alkanes are used as **solvents** because they are **non-polar** and are able to dissolve other non-polar solutes, e.g. hexane and heptane are used as solvents for making fast drying glues and extracting oils from seeds.

Biogas production

Biogas is a **renewable energy source**. It is produced by naturally occurring anaerobic bacteria breaking down organic matter, e.g. manure, in the absence of oxygen in an **anaerobic digester**. The biogas produced is a mixture of approximately **60% methane**, **40% carbon dioxide** and traces of other gases, e.g. hydrogen sulfide.

Biogas can be used directly as a **fuel** for cooking, heating and to generate electricity, or it can be upgraded to almost pure methane, known as **biomethane**, by removal of the other gases.

Uses of alkenes

Alkenes are used as **starting materials** in the manufacture of a wide variety of important chemicals because they readily undergo **addition reactions**. They are used to manufacture ethanol and other **alcohols**, **antifreezes** such as ethane-1,2-diol, **synthetic rubbers** and a variety of **haloalkanes.** Because they can undergo polymerisation reactions, they are used to manufacture a wide range of polymers, also known as **plastics** (see p. 150).

Revision questions

1 Distinguish between a saturated and an unsaturated hydrocarbon.

2 **a** By means of TWO equations, show how ethane and ethene burn in air.
b Would ethane or ethene produce the smokiest flame when burnt in a plentiful supply of oxygen? Explain your answer.

3 Using equations, show the reactions occurring when methane reacts with chlorine in dim light and name the type of reaction occurring.

4 Why are alkenes more reactive than alkanes?

5 Under the correct conditions, ethene reacts with both hydrogen and water. For EACH reaction
a Name the type of reaction involved.
b State the conditions required.
c Name the product formed.
d Write a balanced equation.

6 How would you distinguish between ethane and ethene in the laboratory? Support your answer with an appropriate equation.

7 Give TWO uses of alkanes and TWO uses of alkenes.

Alcohols: $C_nH_{2n+1}OH$ or R–OH

Alcohols (or **alkanols**) have the **hydroxyl group, —OH**, as their functional group. All alcohols undergo similar reactions because they all contain the **hydroxyl group**; however, the strength of the reactions **decreases** as the number of carbon atoms per molecule increases.

General physical properties of alcohols

Alcohol molecules are **polar** because they possess the polar **—OH group**. Because of this:

- Alcohols are **less volatile** than their corresponding alkanes. Because of the polar —OH groups, the forces of attraction between alcohol molecules are stronger than the forces between non-polar alkane molecules with the same number of carbon atoms. All alcohols are **liquids** or **solids** at room temperature. Their boiling points **increase** as the number of carbon atoms per molecule increases.
- Alcohols are **soluble** in water because water is a polar solvent. Their solubility **decreases** as the number of carbon atoms per molecule increases.

Reactions of ethanol

- **Ethanol burns easily in air or oxygen**

  Ethanol burns in air or oxygen to produce **carbon dioxide** and **water** as steam. It burns with a **clear, blue, non-smoky flame** because of the **low** ratio of carbon to hydrogen atoms in the molecules. The reaction is **exothermic**:

$$C_2H_5OH(l) + 3O_2(g) \longrightarrow 2CO_2(g) + 3H_2O(g) \quad \Delta H \text{ –ve}$$

- **Ethanol reacts with sodium**

  Ethanol reacts with sodium to form **sodium ethoxide** (C_2H_5ONa) and **hydrogen**:

$$2C_2H_5OH(l) + 2Na(s) \longrightarrow \underset{\text{sodium ethoxide}}{2C_2H_5ONa(\text{alc sol})} + H_2(g)$$

Note 'alc sol' is 'alcohol solution'

- **Ethanol undergoes dehydration**

  Ethanol can be **dehydrated** to **ethene** in two ways:

  - Heating ethanol at a temperature of about 170 °C with excess concentrated sulfuric acid. The acid acts as a catalyst:

$$C_2H_5OH(l) \xrightarrow[170\ ^\circ C]{\text{conc. } H_2SO_4} C_2H_4(g) + H_2O(g)$$

  - Passing ethanol vapour over heated aluminium oxide. The aluminium oxide acts as a catalyst.

- **Ethanol undergoes oxidation**

  Ethanol is **oxidised** to **ethanoic acid** when heated with acidified potassium manganate(VII) solution or acidified potassium dichromate(VI) solution. Ethanol acts as a **reducing agent**.

$$\underset{\text{ethanol}}{C_2H_5OH(l)} + \underset{\text{from oxidising agent}}{2[O]} \longrightarrow \underset{\text{ethanoic acid}}{CH_3COOH(aq)} + H_2O(l)$$

  Orange acidified potassium dichromate(VI) crystals can be used in the **breathalyser test** to determine the alcohol content of a driver's breath. The driver blows over the crystals and if ethanol vapour is present, it **reduces** the **orange** dichromate(VI) ion ($Cr_2O_7^{2-}$) to the **green** chromium(III) ion (Cr^{3+}). This turns the crystals green.

- **Ethanol reacts with alkanoic acids**

  Ethanol reacts with alkanoic acids to produce an **ester** and **water** (see p. 143).

Production of ethanol by fermentation of carbohydrates

Ethanol can be produced by using **yeast** to **ferment** carbohydrates under **anaerobic** conditions (without oxygen). Yeast produces enzymes that break down complex carbohydrates into simple sugars, mainly glucose. It then produces the enzyme **zymase** which changes the simple sugars into **ethanol** and **carbon dioxide**:

$$\underset{\text{glucose}}{C_6H_{12}O_6(aq)} \xrightarrow{\text{zymase in yeast}} \underset{\text{ethanol}}{2C_2H_5OH(aq)} + 2CO_2(g) \qquad \Delta H \text{ –ve}$$

Fermentation stops when the concentration of ethanol in the fermentation mixture reaches about 14%. At this concentration, the ethanol starts to denature the zymase and this stops it from working. Ethanol which is about 96% pure is then obtained from the fermentation mixture using **fractional distillation** (see p. 14), collecting the fraction that distils at 78 °C.

Fermentation of carbohydrates is used to produce a variety of different **alcoholic beverages**, including wine and rum.

- **Wine** is mainly made from **grapes** in a winery. The yeast is added to the crushed grapes and it ferments the sugars present. Air should not come into contact with the wine because certain aerobic bacteria **oxidise** the ethanol to **ethanoic acid**, or vinegar, causing the wine to become sour:

$$C_2H_5OH(aq) + O_2(g) \xrightarrow{\text{aerobic bacteria}} \underset{\text{ethanoic acid}}{CH_3COOH(aq)} + H_2O(l)$$

- **Rum** is made from **molasses** in a rum distillery. The yeast is added to the molasses and it ferments the sugars present. The mixture is then **fractionally distilled**. The distillate is diluted with water and transferred to oak barrels to be aged.

Alkanoic acids: $C_nH_{2n+1}COOH$ or R–COOH

Alkanoic acids (or **carboxylic acids**) have the **carboxyl** group, **—COOH**, as their functional group.

Alkanoic acid molecules are **polar** due to the **—OH** part of the functional group being polar. Like alcohols, alkanoic acids are **less volatile** than their corresponding alkanes and are **soluble** in water. When they dissolve in water they **partially** ionise, and are therefore **weak acids**:

e.g. $CH_3COOH(aq) \rightleftharpoons CH_3COO^-(aq) + H^+(aq)$

ethanoic acid → ethanoate ion

Reactions of aqueous ethanoic acid

Aqueous solutions of alkanoic acids react in a similar way to inorganic acids, such as hydrochloric acid, though the reactions are **slower** because the acids are weak. The hydrogen in the functional group can be replaced directly or indirectly by a metal to form a salt. The salts formed by ethanoic acid are called **ethanoates**.

- **Aqueous ethanoic acid reacts with reactive metals**

  Ethanoic acid reacts with reactive metals to form a **salt** and **hydrogen** gas:

  e.g. $Ca(s) + 2CH_3COOH(aq) \longrightarrow (CH_3COO)_2Ca(aq) + H_2(g)$

  calcium ethanoate

- **Aqueous ethanoic acid reacts with metal oxides and metal hydroxides**

  Ethanoic acid reacts with metal oxides and hydroxides to form a **salt** and **water**:

  e.g. $MgO(s) + 2CH_3COOH(aq) \longrightarrow (CH_3COO)_2Mg(aq) + H_2O(l)$

  magnesium ethanoate

  $KOH(aq) + CH_3COOH(aq) \longrightarrow CH_3COOK(aq) + H_2O(l)$

  potassium ethanoate

- **Aqueous ethanoic acid reacts with metal carbonates**

  Ethanoic acid reacts with metal carbonates to form a **salt**, **carbon dioxide** and **water**.

  e.g. $CuCO_3(s) + 2CH_3COOH(aq) \longrightarrow (CH_3COO)_2Cu(aq) + CO_2(g) + H_2O(l)$

  copper(II) ethanoate

Reactions of anhydrous ethanoic acid

Anhydrous (water-free) ethanoic acid reacts with **alcohols** to produce an **ester** and **water** (see below).

Esters: $C_nH_{2n+1}COOC_xH_{2x+1}$ or R–COO–R′

An **ester** is formed when an **alkanoic acid** reacts with an **alcohol**. The reaction is a type of **condensation reaction** where the two molecules join to form a larger molecule with the loss of a water molecule. This particular condensation reaction is known as **esterification**, and it requires concentrated sulfuric acid and heat.

$$\textbf{alkanoic acid} + \textbf{alcohol} \underset{\text{heat}}{\overset{\text{conc. } H_2SO_4}{\rightleftharpoons}} \textbf{ester} + \textbf{water}$$

Example

$$H{-}\underset{H}{\overset{H}{C}}{-}C{\overset{O}{\underset{O-H}{}}} + H{-}O{-}\underset{H}{\overset{H}{C}}{-}\underset{H}{\overset{H}{C}}{-}\underset{H}{\overset{H}{C}}{-}H \xrightleftharpoons[\text{heat}]{\text{conc. } H_2SO_4} H{-}\underset{H}{\overset{H}{C}}{-}\overset{O}{\overset{\|}{C}}{-}O{-}\underset{H}{\overset{H}{C}}{-}\underset{H}{\overset{H}{C}}{-}\underset{H}{\overset{H}{C}}{-}H + H_2O$$

H_2O lost

ethanoic acid **propanol** **propyl ethanoate** **water**

or $$CH_3COOH(l) + C_3H_7OH(l) \xrightleftharpoons[\text{heat}]{\text{conc. } H_2SO_4} CH_3COOC_3H_7(l) + H_2O(l)$$

The concentrated sulfuric acid is added for **two** reasons:

- To act as a **catalyst** to **speed up** the reaction.
- To **remove the water** produced during the reaction. This favours the forward reaction and **increases the yield** of ester.

Esters have the **ester group, —COO—**, as their functional group:

$$-\overset{O}{\overset{\|}{C}}-O-$$

Many esters are found naturally. Those with a low number of carbon atoms in their molecules are volatile liquids which usually have very distinctive sweet, fruity smells. They are found especially in flowers and fruit. Animal fats and vegetable oils are esters formed from **fatty acids** (long chain alkanoic acids), and **glycerol** (an alcohol).

Writing the formulae and names of esters

When writing formula and name of an ester:

- **Formula:** the part from the acid comes first
  the part from the **alcohol** comes second
- **Name:** the prefix from the **alcohol**, ending in '-yl', comes first
  the prefix from the acid, ending in '-anoate', comes second

Example

$$C_3H_7COOH + CH_3OH \rightleftharpoons C_3H_7COOCH_3 + H_2O$$

butanoic acid methanol **methyl** butanoate

Table 16.3 *Formulae and names of some esters*

Alkanoic acid	Alcohol	Ester
CH_3COOH ethanoic acid	C_4H_9OH butanol	$CH_3COOC_4H_9$ **butyl** ethanoate
HCOOH methanoic acid	C_2H_5OH ethanol	$HCOOC_2H_5$ **ethyl** methanoate
C_2H_5COOH propanoic acid	C_3H_7OH propanol	$C_2H_5COOC_3H_7$ **propyl** propanoate

Hydrolysis of esters

During **hydrolysis**, molecules of a compound are broken down into smaller molecules by reacting the compound with **water**. Hydrolysis of esters can be carried out by heating the ester with a dilute acid or an aqueous alkali.

- **Acid hydrolysis** involves heating the ester with dilute hydrochloric or sulfuric acid. The acid acts as a **catalyst**. The products are the **alkanoic acid** and the **alcohol** from which the ester was formed.

e.g. $$\underset{\text{propyl ethanoate}}{CH_3COOC_3H_7(l)} + H_2O(l) \xrightleftharpoons{H^+ \text{ ions}} \underset{\text{ethanoic acid}}{CH_3COOH(aq)} + \underset{\text{propanol}}{C_3H_7OH(aq)}$$

- **Alkaline hydrolysis** involves heating the ester with potassium or sodium hydroxide solution. The products are the **potassium** or **sodium salt** of the **alkanoic acid** and the **alcohol** from which the ester was formed.

e.g. $$\underset{\text{propyl ethanoate}}{CH_3COOC_3H_7(l)} + NaOH(aq) \longrightarrow \underset{\text{sodium ethanoate}}{CH_3COONa(aq)} + \underset{\text{propanol}}{C_3H_7OH(aq)}$$

Saponification of fats and oils – making soap

Saponification refers to the process that produces **soap**. During saponification, large ester molecules found in animal fats and vegetable oils undergo **alkaline hydrolysis** by being boiled with concentrated potassium or sodium hydroxide solution. The reaction produces **soap**, which is the potassium or sodium salt of a long chain alkanoic acid (**fatty acid**) and **glycerol** ($C_3H_5(OH)_3$).

Example

The fat **glyceryl octadecanoate** ($(C_{17}H_{35}COO)_3C_3H_5$) is an ester of **octadecanoic acid** ($C_{17}H_{35}COOH$) and **glycerol** ($C_3H_5(OH)_3$). Saponification of glyceryl octadecanoate by boiling with sodium hydroxide solution, forms **sodium octadecanoate** ($C_{17}H_{35}COONa$) and **glycerol**. Sodium octadecanoate, also called sodium stearate, is the most common form of **soap**.

$$\underset{\substack{\text{glyceryl octadecanoate}\\\text{(fat)}}}{(C_{17}H_{35}COO)_3C_3H_5(l)} + 3NaOH(aq) \xrightarrow{\text{heat}} \underset{\substack{\text{sodium octadecanoate}\\\text{(soap)}}}{3C_{17}H_{35}COONa(aq)} + \underset{\text{glycerol}}{C_3H_5(OH)_3(aq)}$$

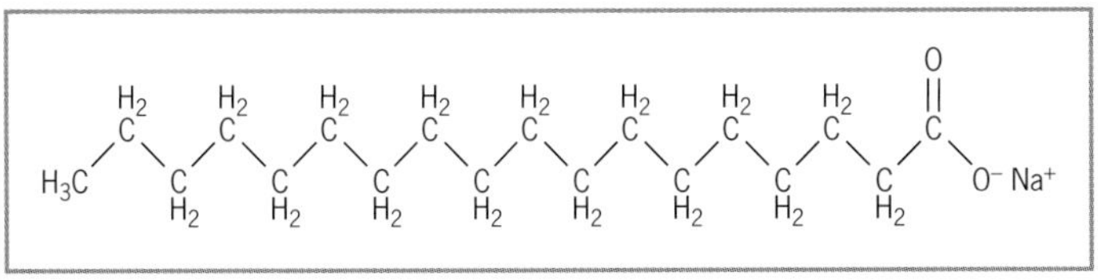

Figure 16.2 *A soap molecule*

Soapy and soapless detergents

Soapy and **soapless detergents** are substances which are added to water to remove dirt, e.g. from the skin, clothes, household surfaces and floors.

- **Soapy detergents** are made by boiling animal fats or vegetable fats and oils with concentrated potassium or sodium hydroxide solution. They may be simply called **soaps**, e.g. sodium octadecanoate, $C_{17}H_{35}COO^-Na^+$.
- **Soapless detergents** are formed from **petroleum**. They are also known as 'synthetic detergents' and may be simply called **detergents**, e.g. sodium dodecyl sulfate, $C_{12}H_{25}OSO_3^-Na^+$.

Table 16.4 *Comparing soapy and soapless detergents*

Soapy detergents	Soapless detergents
Manufactured from fats and oils, a **renewable** resource which will not run out.	Manufactured from petroleum, a **non-renewable** resource which will eventually run out.
Do not lather easily in hard water containing Ca^{2+} and Mg^{2+} ions. Their calcium and magnesium salts are insoluble and form unpleasant **scum** (see p. 180). This wastes soap, discolours clothes and forms an unpleasant grey, greasy layer around sinks, baths and showers.	Lather easily in hard water. Their calcium and magnesium salts are soluble so they **do not** form scum.
Are **biodegradable**, meaning they are broken down by bacteria in the environment. They **do not** cause foam to form on waterways such as lakes and rivers, or in sewage systems.	Some are **non-biodegradable**. These can cause **foam** to form on waterways and in sewage systems. This causes aquatic organisms to die because oxygen can no longer dissolve in the water, and it makes sewage treatment difficult. Most modern soapless detergents are now biodegradable.
Do not contain phosphates, so they **do not** cause eutrophication of aquatic environments.	Some contain **phosphates**, added to improve their cleaning ability. Phosphates cause pollution of aquatic environments by causing the rapid growth of green algae known as **eutrophication** (see Table 21.6, p. 177).

Foam on a river

Eutrophication

Revision questions

8 Give the name and formula of the functional group in EACH of the following:

a an alcohol **b** an alkanoic acid **c** an ester

9 Explain why ethanol is:

a less volatile than ethane **b** soluble in water.

10 **a** What is produced when ethanol is heated with excess concentrated sulfuric acid?

b Name the type of reaction occurring in **a** above.

11 Explain the chemical principles of the breathalyser test.

12 **a** What is the type of reaction that produces ethanol from carbohydrates called?

b What else is needed to carry out the reaction named in **a** above?

c Why must air not be allowed to come into contact with wine as it is being made?

13 Write a balanced equation for the reaction between aqueous ethanoic acid and EACH of the following:

a aluminium **b** calcium hydroxide **c** zinc oxide **d** magnesium carbonate

14 **a** Write an equation for the reaction between ethanoic acid and ethanol and state the conditions required.

b Name the type of reaction occurring in **a** above.

15 Name EACH of the following esters:

a $C_2H_5COOC_2H_5$ **b** $HCOOC_4H_9$ **c** $C_3H_7COOCH_3$ **d** $CH_3COOC_3H_7$

16 Give the formula of EACH of the following esters:

a propyl butanoate **b** methyl propanoate

c butyl ethanoate **d** ethyl methanoate

17 What happens during a saponification reaction?

18 Give THREE differences between a soapy and a soapless detergent.

17 Polymers

Polymers are large molecules (macromolecules) which are made from many repeating small molecules, known as **monomers**, bonded together. Some occur in nature, whilst others are man-made.

*A **polymer** is a macromolecule formed by linking together 50 or more monomers, usually in chains.*

Formation of polymers

A polymer may contain thousands of monomers. The reaction by which a polymer is formed from monomers is known as **polymerisation**. There are **two** types:

- **Addition polymerisation** occurs when **unsaturated** monomers, such as **alkenes**, are linked together to form a **saturated** polymer known as an **addition polymer**. One bond in the double bond of each alkene molecule breaks and the molecules bond to one another in chains by single covalent bonds between adjacent carbon atoms.
- **Condensation polymerisation** involves linking monomers together in chains by eliminating a **small molecule**, usually water, from between adjacent monomers. For this to occur, the monomer molecules must have **two functional groups**. The polymers formed are known as **condensation polymers**.

Table 17.1 *Addition and condensation polymerisation compared*

Addition polymerisation	Condensation polymerisation
The polymer is the **only product**.	**Two products** are formed, the polymer and small molecules, usually water.
The polymer is made from monomers which are all the **same**.	The polymer is often made from **more than one** type of monomer.
The **empirical formula** of the polymer is the **same** as the monomer from which it was formed; no atoms are lost during its formation.	The **empirical formula** of the polymer is **different** from the monomers; atoms are lost during its formation.

Table 17.2 *Classification of polymers*

Type of polymerisation	Type of polymer	Examples	
		Synthetic (man-made)	Natural
Addition polymerisation	Polyalkene	Polyethene Polypropene Polychloroethene (polyvinyl chloride or PVC) Polystyrene (styrofoam)	None exist
Condensation polymerisation	Polyester	Polyethylene terephthalate (PET, Terylene or Dacron)	None exist
	Polyamide	Nylon	Proteins
	Polysaccharide	None exist	Starch Glycogen (animal starch) Cellulose

Table 17.3 *Formation of the different polymers*

Type of polymer	Type of linkage between monomers	Structure and name(s) of the monomer(s)	Structure and name of the polymer
Polyalkene	Alkane 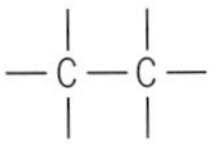 —C—C—	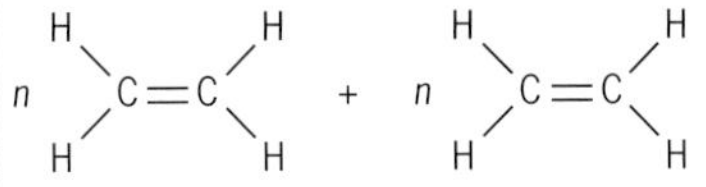 n $H_2C{=}CH_2$ + n $H_2C{=}CH_2$ **ethene** **ethene**	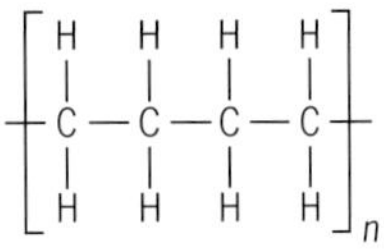 $-[CH_2-CH_2-CH_2-CH_2]_n-$ **polyethene**
		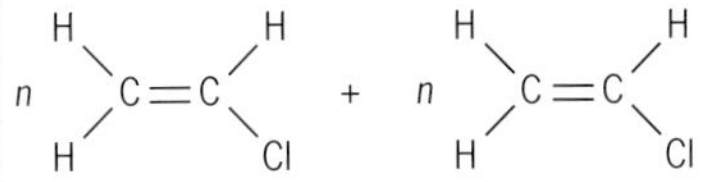 n $H_2C{=}CHCl$ + n $H_2C{=}CHCl$ **chloroethene** **chloroethene**	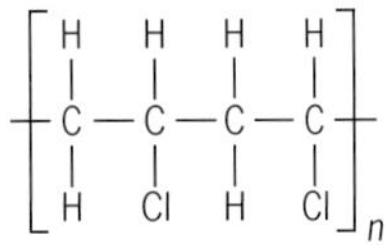 $-[CH_2-CHCl-CH_2-CHCl]_n-$ **polychloroethene** or **polyvinyl chloride (PVC)**
		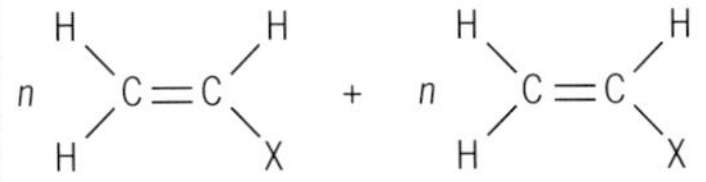 n $H_2C{=}CHX$ + n $H_2C{=}CHX$ other **alkene** molecules where X represents another atom, hydrocarbon group or functional group	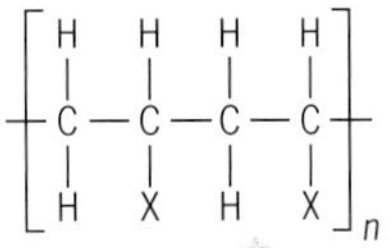 $-[CH_2-CHX-CH_2-CHX]_n-$ If X = **CH_3**, the polymer is **polypropene** If X = **C_6H_5**, the polymer is **polystyrene**
Polyester	Ester 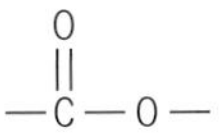 —C(=O)—O— or — COO —	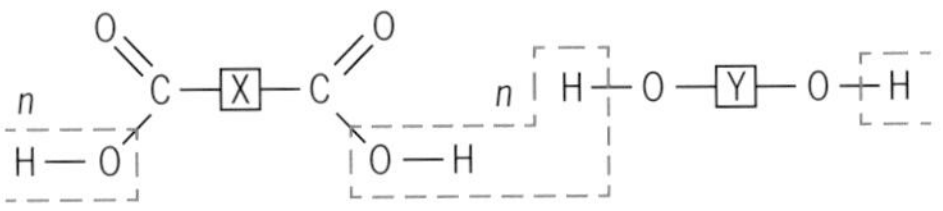 n H—O—C(=O)—[X]—C(=O)—O—H + n H—O—[Y]—O—H **diacid** **dialcohol** [Y] and [Y] represent variable hydrocarbon groups	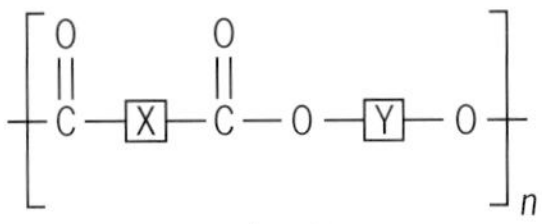 $-[$C(=O)—[X]—C(=O)—O—[Y]—O$]_n-$ **a polyester**
Polyamide	Amide (peptide) 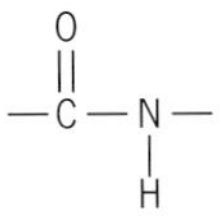 —C(=O)—N(H)— or — CONH —	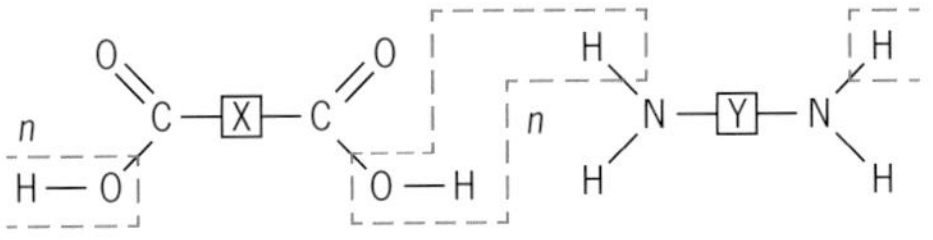 n H—O—C(=O)—[X]—C(=O)—O—H + n H₂N—[Y]—NH₂ **diacid** **diamine** [Y] and [Y] represent variable hydrocarbon groups	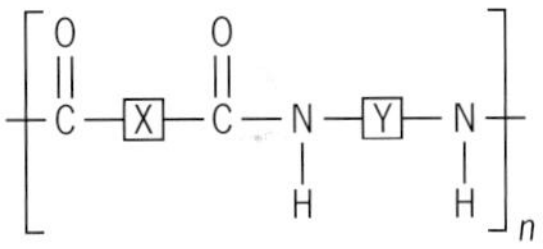 $-[$C(=O)—[X]—C(=O)—N(H)—[Y]—N(H)$]_n-$ **a polyamide**
		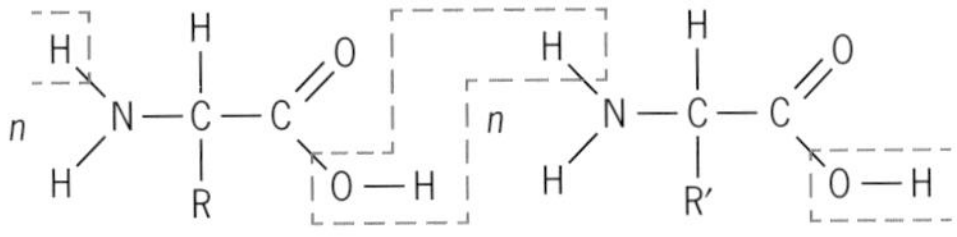 n H₂N—CH(R)—C(=O)—O—H + n H₂N—CH(R′)—C(=O)—O—H **amino acid** **amino acid** R and R′ represent different hydrocarbon groups which may also contain oxygen, sulfur and nitrogen atoms. There are 20 of these groups, so there are 20 different amino acids.	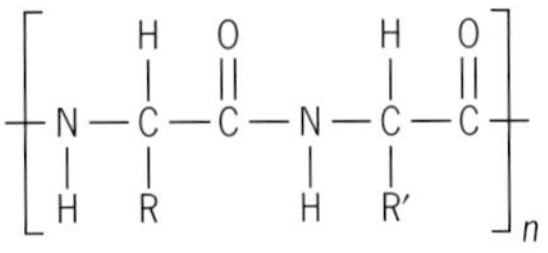 $-[$N(H)—CH(R)—C(=O)—N(H)—CH(R′)—C(=O)$]_n-$ **a protein**
Polysaccharide	Ether  —O—	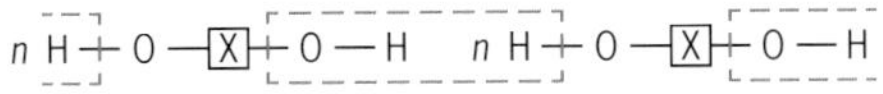 n H—O—[X]—O—H n H—O—[X]—O—H **monosaccharide** ($C_6H_{12}O_6$) **monosaccharide** ($C_6H_{12}O_6$) e.g. glucose, where [X] represents $C_6H_{10}O_4$	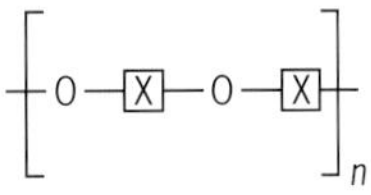 $-[$O—[X]—O—[X]$]_n-$ **a polysaccharide**

Note —O—H H— represents the loss of a water molecule (H_2O) from between adjacent monomers.

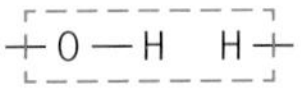

Uses of polymers

Synthetic polymers, commonly known as **plastics**, have a great many uses because their properties make them superior to many other materials.

- They are durable, i.e. they are resistant to physical and chemical damage, and biological decay.
- They are easily moulded into many different shapes.
- They are light but strong.
- They are easily dyed different colours.
- They can be made to be rigid or flexible.
- They are good thermal and electrical insulators.
- They can be easily welded or joined.
- They can be spun into fibres because their molecules are extremely long.

Table 17.4 *Uses of polymers*

Polymer type	Name of polymer	Uses of polymer
Polyalkene	Polyethene	To make plastic bags, plastic bottles, washing up bowls, buckets, packaging for food and cling wrap.
	Polychloroethene (PVC)	To make water and sewer pipes, guttering, window and door frames, and insulation for electrical wires and cables.
	Polypropene	To make ropes, carpets, plastic toys, plastic food containers and furniture.
	Polystyrene (styrofoam)	To make containers for fast food and drinks, packaging materials and insulation materials.
Polyester	Polyethylene terephthalate (PET, Terylene or Dacron)	To make PET bottles for soft drinks. Fibres are used to make clothing, boat sails, carpets and fibre filling for winter clothing, sleeping bags and pillows.
Polyamide	Nylon	Fibres are used to make fishing lines and nets, ropes, carpets, parachutes, tents and clothing, especially if stretch is required, e.g. sports wear and nylon stockings.
	Protein	To build body cells, hair and nails. To make enzymes and antibodies.
Polysaccharide	Starch and glycogen	Stored as food reserves in living organisms. When needed, they can be broken down into glucose which is used in respiration to produce energy.
	Cellulose	To build plant cells walls.

Impact of synthetic polymers (plastics) on the environment

Synthetic polymers can have **harmful effects** on living organisms and the environment:

- Plastics are made from a **non-renewable resource** (petroleum). Their manufacture is contributing to the depletion of petroleum world-wide.
- Most plastics are **non-biodegradable.** Waste plastics build up in the environment causing pollution of land and water.
- Plastics are directly harmful to **aquatic organisms** such as sea turtles, due to ingestion, entanglement and suffocation.
- Various **toxic chemicals** are released into the environment during the manufacture of plastics, and some of these continue to be released from the plastic items during use and when discarded.

- Many plastics are **flammable**, therefore they pose fire hazards.
- When burnt, plastics produce **dense smoke** and **poisonous gases** which can lead to air pollution.

Revision questions

1 What is a polymer?

2 Give THREE differences between addition polymerisation and condensation polymerisation.

3 Draw the partial structure of a polymer which could be formed from EACH of the following. In EACH case name the type of polymer formed and the type of linkage it contains.

a

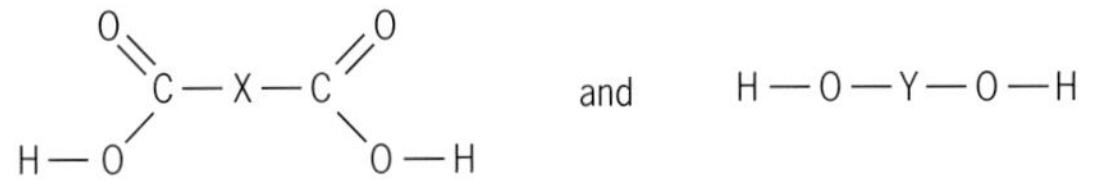

```
   O          O
    \\      //
     C — X — C                and      H — O — Y — O — H
    /        \
H — O         O — H
```

b

```
   O          O                      H           H
    \\      //                        \         /
     C — X — C                and      N — Y — N
    /        \                        /         \
H — O         O — H                  H           H
```

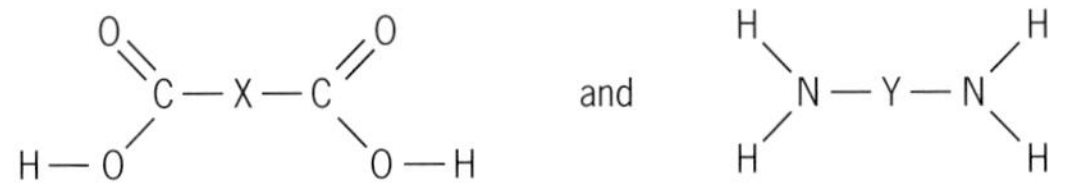

c

```
H       H
 \     /
  C = C
 /     \
H       C6H5
```

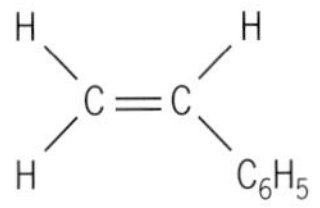

4 Draw the partial structure of a polysaccharide molecule composed of THREE monosaccharide molecules.

5 Give TWO uses of EACH of the following:

a a <u>named</u> polyalkene

b a <u>named</u> polyester

c a <u>named</u> natural polyamide

6 Give ONE use of EACH of the following:

a a <u>named</u> synthetic polyamide

b a <u>named</u> polysaccharide

Exam-style questions – Chapters 14 to 17

Structured questions

1 **a)** Petroleum (crude oil) is a complex mixture of hydrocarbons.

i) Name the process by which petroleum is separated into its various fractions. **(1 mark)**

ii) Give ONE major use in EACH case of any of TWO named fractions obtained from petroleum. A different use must be given for EACH fraction. **(2 marks)**

b) After petroleum is separated into its various fractions, ethene is obtained by a process known as cracking.

i) What is meant by the term 'cracking'? **(1 mark)**

ii) Write an equation to show how ethene can be obtained by cracking C_7H_{16} and name the other product of the cracking process. **(2 marks)**

iii) Ethene reacts with chlorine under standard laboratory conditions. Write an equation to summarise this reaction. **(1 mark)**

c) Under the right conditions, ethene can be converted to the polymer, polyethene.

i) Define the term 'polymer'. **(1 mark)**

ii) Identify the type of polymerisation used in the manufacture of polyethene. **(1 mark)**

iii) Draw the partial structure of the polymer that would be obtained by polymerising THREE ethene molecules. **(1 mark)**

iv) Give TWO uses of polyethene. **(2 marks)**

d) i) On analysis, a compound was found to contain 85.7% carbon and 14.3% hydrogen. Determine the empirical formula of this compound.
(Relative atomic masses: H = 1; C = 12) **(2 marks)**

ii) To which homologous series does this compound belong? **(1 mark)**

Total 15 marks

2 The structures of FOUR organic compounds, A, B, C and D, are given in Figure 1.

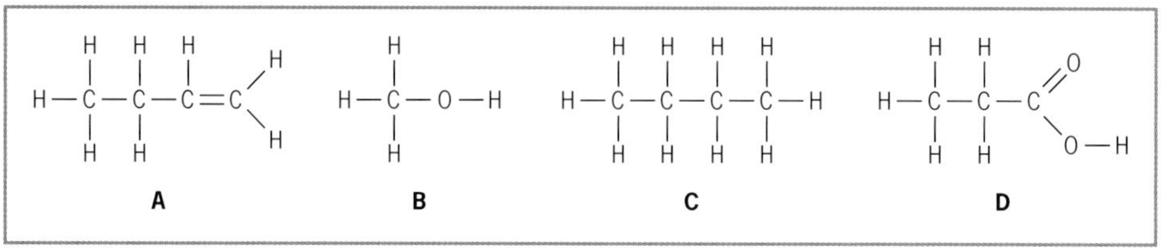

```
    H  H  H                H            H  H  H  H              H  H
    |  |  |      H         |            |  |  |  |              |  |     O
H—C—C—C=C/            H—C—O—H       H—C—C—C—C—H           H—C—C—C//
    |  |       \H          |            |  |  |  |              |  |  \
    H  H                   H            H  H  H  H              H  H   O—H
       A                   B                 C                     D
```

Figure 1 *Structure of four organic compounds*

a) Name the homologous series to which B belongs and give a reason for your answer. **(2 marks)**

b) i) Define the term 'structural isomerism'. **(1 mark)**

ii) Draw the fully displayed structural formula of ONE isomer of A and name the isomer. **(2 marks)**

c) Describe ONE chemical test that could be used in the laboratory to distinguish between A and C. **(2 marks)**

d) Both A and C burn in oxygen.

i) Write a balanced chemical equation for the reaction occurring when C burns in excess oxygen. **(2 marks)**

ii) If both A and C are burnt in excess oxygen, which compound would produce the cleanest flame? Give a reason for your answer. **(2 marks)**

e) Under the correct conditions compounds B and D react.

i) Identify the conditions needed for the reaction. **(1 mark)**

ii) Write an equation for the reaction and name the organic product. **(2 marks)**

iii) Name the type of reaction occurring. **(1 mark)**

Total 15 marks

3 Figure 2 shows the reactions of some organic compounds. Use the information in Figure 2 to answer the questions that follow. Compound E has been identified for you and information is given about some of the other compounds.

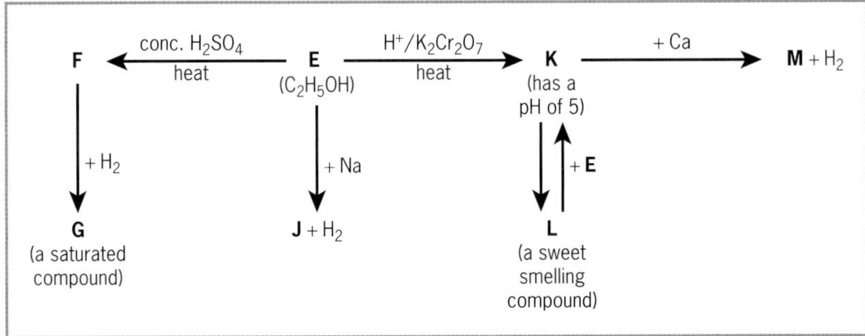

Figure 2 *Reactions of some organic compounds*

a) Draw the fully displayed structural formula of EACH of the compounds F, K and L and name EACH compound. **(6 marks)**

b) Name the type of reaction involved in converting:

i) E to F. **(1 mark)**

ii) F to G. **(1 mark)**

iii) E to K. **(1 mark)**

c) Identify the reaction conditions needed to convert F to G. **(1 mark)**

d) Write a balanced equation for EACH of the following:

i) The reaction between E and sodium. **(2 marks)**

ii) The reaction between K and calcium. **(2 marks)**

e) The reaction occurring in the conversion of E to K is applied in the breathalyser test. What is this test used for? **(1 mark)**

Total 15 marks

Extended-response questions

4 **a)** Organic compounds can be classified into homologous series.

i) Give TWO characteristics of a homologous series. **(2 marks)**

ii) Methane is a member of the homologous series of alkanes. Show, by means of a dot and cross diagram, how the atoms are bonded in a methane molecule. **(2 marks)**

iii) Give TWO reasons why alkanes are used extensively as fuels. **(2 marks)**

b) Ethanol is a member of the alcohol series.

i) Describe how you would prepare a sample of about 12% ethanol in the laboratory from a suitable carbohydrate source. Your answer must include a relevant equation. **(3 marks)**

ii) Draw a labelled diagram of the apparatus you would use to obtain a sample of almost pure ethanol from the mixture obtained in **i)** above. **(4 marks)**

iii) Susan, who has been making wine at home for years, noticed that her latest batch had gone sour. Explain, with an appropriate equation, how this happened. **(2 marks)**

Total 15 marks

5 **a)** A compound X, which has the formula shown below, is heated with dilute hydrochloric acid.

$$CH_3CH_2CH_2-\overset{\overset{\displaystyle O}{\|}}{C}-O-CH_3$$

i) Identify the homologous series to which X belongs, give the names and formulae of the products of the reaction and name the type of reaction occurring. **(4 marks)**

ii) What difference, if any, would there have been in the products of the reaction if X had been heated with sodium hydroxide solution instead of hydrochloric acid? **(2 marks)**

b) One use of vegetable oils and animal fats is to manufacture soap.

i) How is soap produced from the fat glyceryl octadecanoate? Give an equation for the reaction and name the process involved. **(4 marks)**

ii) Glycerol, a by-product in the manufacture of soap, has the formula $C_3H_5(OH)_3$. Would you expect glycerol to be soluble or insoluble in water? Explain your answer. **(2 marks)**

iii) Soap is one type of detergent used to remove dirt from clothes. The other type is known as a soapless detergent. As a citizen concerned about the environment, would you choose to use soapy or soapless detergents on a regular basis? Give THREE reasons to support your choice. **(3 marks)**

Total 15 marks

18 Characteristics of metals

Metals are elements that are found mainly in Groups I, II and III of the periodic table, and between Groups II and III as the transition metals. The atoms of most metals have 1, 2 or 3 valence electrons.

Physical properties of metals

Metals have common **physical properties** as a result of the **metallic bonding** between metal atoms in the metallic lattice (see p. 39).

- Metals have **high** melting and boiling points.
- Metals are **solid** at room temperature (except mercury).
- Metals are **good** conductors of electricity and heat.
- Metals are **malleable** and **ductile**.
- Metals are **shiny** in appearance.
- Metals have **high** densities.

Chemical properties of metals

When they react, metals ionise by **losing** electrons to form positive **cations**. The metal behaves as a **reducing agent** because it gives electrons to the other reactant (it causes the other reactant to **gain** electrons). This results in the formation of **ionic compounds**.

$$M \longrightarrow M^{n+} + ne^-$$

Some metals react vigorously, even violently, with acids, oxygen and water, whilst others are relatively unreactive. Potassium, sodium, calcium and magnesium are the most reactive, whilst aluminium, zinc and iron are less reactive, and copper and silver are relatively unreactive.

Reactions of metals with dilute acids

When a metal reacts with dilute hydrochloric or sulfuric acid, it forms a **salt** and **hydrogen**:

metal + acid ⟶ salt + hydrogen

e.g.

$$2Na(s) + 2HCl(aq) \longrightarrow 2NaCl(aq) + H_2(g)$$

$$2Al(s) + 3H_2SO_4(aq) \longrightarrow Al_2(SO_4)_3(aq) + 3H_2(g)$$

Reactions of metals with oxygen

When a metal reacts with oxygen, it forms a **metal oxide**. Aluminium, zinc and lead(II) oxide are **amphoteric**, the oxides of the other metals are **basic**:

metal + oxygen ⟶ metal oxide

e.g.

$$4K(s) + O_2(g) \longrightarrow 2K_2O(s)$$

$$2Cu(s) + O_2(g) \longrightarrow 2CuO(s)$$

Reactions of metals with water

- When a metal reacts with **water**, it forms a **metal hydroxide** and **hydrogen**:

metal + water ⟶ metal hydroxide + hydrogen

e.g.

$$Ca(s) + 2H_2O(l) \longrightarrow Ca(OH)_2(aq) + H_2(g)$$

- When a metal reacts with **steam**, it forms a **metal oxide** and **hydrogen**:

metal + steam ⟶ metal oxide + hydrogen

e.g. $3Fe(s) + 4H_2O(g) \longrightarrow Fe_3O_4(s) + 4H_2(g)$
iron(II, III) oxide

Table 18.1 *Summary of reactions of metals with dilute acids, oxygen and water*

Metal	Description of the reaction with dilute acids	Description of the reaction when the metal is heated in air	Description of the reaction with water
Potassium	Reacts extremely violently.	Burns vigorously with a lilac flame.	Reacts very vigorously with cold water.
Sodium	Reacts violently.	Burns vigorously with an orange flame.	Reacts vigorously with cold water.
Calcium	Reacts fairly violently.	Burns very easily with a brick red flame.	Reacts moderately with cold water.
Magnesium	Reacts very vigorously.	Burns easily with a bright white flame.	Reacts very slowly with cold water, slowly with hot water. Reacts vigorously with steam.
Aluminium	Reacts vigorously.	Burn when heated strongly, especially if powdered.	Do not react with cold water or hot water. React with steam.
Zinc	Reacts fairly vigorously.		
Iron	Reacts very slowly.		
Copper	Do not react with dilute acids.	Does not burn when heated, but forms an oxide coating if heated very strongly.	Do not react with water or steam.
Silver		Does not react, even when heated very strongly.	

Reactions of metal compounds

All metal oxides, hydroxides and carbonates react with **dilute acids** and many metal hydroxides, carbonates and nitrates decompose when heated.

Reactions of metal compounds with dilute acids

- **Metal oxides** react with acids to form a **salt** and **water**:

metal oxide + acid ⟶ salt + water

e.g. $ZnO(s) + 2HCl(aq) \longrightarrow ZnCl_2(aq) + H_2O(l)$

- **Metal hydroxides** react with acids to form a **salt** and **water**:

metal hydroxide + acid ⟶ salt + water

e.g. $Mg(OH)_2(s) + H_2SO_4(aq) \longrightarrow MgSO_4(aq) + 2H_2O(l)$

- **Metal carbonates** react with acids to form a **salt**, **carbon dioxide** and **water**:

metal carbonate + acid ⟶ salt + carbon dioxide + water

e.g. $CuCO_3(s) + 2HNO_3(aq) \longrightarrow Cu(NO_3)_2(aq) + CO_2(g) + H_2O(l)$

Decomposition of metal compounds when heated

Compounds of potassium and sodium are **stable** and do not decompose on heating or, in the case of the nitrates, decompose only slightly. Compounds of other metals decompose when heated and the ease of decomposition increases going **down** the reactivity series (see p. 158).

Table 18.2 *The effect of heat on metal carbonates, hydroxides and nitrates*

Metal	Metal compounds		
	Hydroxides	Carbonates	Nitrates
Potassium Sodium	The hydroxides are stable; they **do not** decompose.	The carbonates are stable; they **do not** decompose.	Decompose slightly to form the **metal nitrite** and **oxygen**: e.g. $2NaNO_3(s) \longrightarrow 2NaNO_2(s) + O_2(g)$
Calcium Magnesium Aluminium Zinc Iron Lead Copper	Decompose to form the **metal oxide** and **water** (as steam): e.g. $Ca(OH)_2(s) \longrightarrow CaO(s) + H_2O(g)$	Decompose to form the **metal oxide** and **carbon dioxide**: e.g. $CaCO_3(s) \longrightarrow CaO(s) + CO_2(g)$	Decompose to form the **metal oxide**, **nitrogen dioxide** and **oxygen**: e.g. $2Ca(NO_3)_2(s) \longrightarrow 2CaO(s) + 4NO_2(g) + O_2(g)$
Silver	Silver hydroxide is too unstable to exist.	Silver carbonate is too unstable to exist.	Decomposes to form **silver**, **nitrogen dioxide** and **oxygen**: $2AgNO_3(s) \longrightarrow 2Ag(s) + 2NO_2(g) + O_2(g)$

Revision questions

1. Give FOUR physical properties that are typical of metals.
2. Explain why metals behave as reducing agents in their reactions.
3. Describe THREE different reactions which are typical of metals.
4. Write a balanced equation for EACH of the following reactions and name the products in EACH case:
   a zinc reacting with hydrochloric acid
   b sodium reacting with cold water
   c magnesium reacting with steam
   d aluminium reacting with oxygen
5. Certain compounds of metals react with dilute acids. Using THREE different compounds of calcium, write an equations to show how EACH compound reacts with dilute hydrochloric acid.
6. What effect, if any, would heat have on EACH of the compounds listed below? Write balanced equations for the reactions where appropriate.
   a copper(II) carbonate
   b lead(II) nitrate
   c sodium carbonate
   d magnesium hydroxide
   e potassium nitrate
   f sodium hydroxide

19 The reactivity, extraction and uses of metals

The reactivity of metals varies greatly and this has an impact on how the metals occur in the Earth's crust, how they are extracted and their uses. When metals are arranged in order according to how reactive they are, the **reactivity series of metals** is produced.

The reactivity series of metals

The **reactivity series of metals** arranges the metals in order from the most reactive to the least reactive. It is based on the following:

- How vigorously the metals react with **dilute acids** (hydrochloric acid and sulfuric acid), **oxygen** and **water** (see Table 18.1, p. 156).
- How easily metal compounds are **decomposed** when they are heated (see Table 18.2, p. 157).
- Whether or not a metal will **displace** another metal from its compounds (see p. 159).

Table 19.1 *The reactivity series of metals*

Metal	Symbol	Reactivity
potassium	K	most reactive ↑
sodium	Na	
calcium	Ca	
magnesium	Mg	
aluminium	Al	**Increasing**:
zinc	Zn	• ease of ionisation
iron	Fe	• reactivity
lead	Pb	• strength as a reducing agent
(hydrogen)	(H)	• stability of compounds
copper	Cu	
mercury	Hg	
silver	Ag	
gold	Au	least reactive

The reactivity is determined by how easily the metal atoms **ionise**; the more easily a metal ionises, the more reactive it is.

- Metals at the **top** of the series ionise the most easily, which makes them the **most reactive**. They are the **strongest reducing agents** because they give away electrons the most easily. Their ions are very **stable**, which makes their compounds are very stable.
- Metals at the **bottom** of the series ionise the least easily which makes them the **least reactive**. They are the **weakest reducing agents** because they give away electrons the least easily. Their ions are very **unstable**, which makes their compounds very unstable.

Displacement reactions

A **displacement reaction** occurs when a metal in its free state takes the place of another metal in a compound. A **more reactive metal** always displaces a **less reactive metal** from its compounds. Atoms of the more reactive metal **ionise** to form ions and the ions of the less reactive metal are **discharged** to form atoms.

$$\mathbf{C} + \mathbf{DX} \longrightarrow \mathbf{CX} + \mathbf{D}$$

C: higher in the series than D

DX: lower in the series than C

- C is higher in the reactivity series, therefore is **more** reactive than D. C **ionises** to form C^{n+} ions:

$$C \longrightarrow C^{n+} + ne^-$$

- D is lower in the reactivity series, therefore is **less** reactive than C. The D^{n+} ions are **discharged** to form D atoms:

$$D^{n+} + ne^- \longrightarrow D$$

These are also **redox reactions**. The metal in its free state is a **stronger** reducing agent than the metal in the compound.

e.g. $Zn(s) + CuSO_4(aq) \longrightarrow ZnSO_4(aq) + Cu(s)$

Ionically: $Zn(s) + Cu^{2+}(aq) \longrightarrow Zn^{2+}(aq) + Cu(s)$

The extraction of metals from their ores

Silver, gold and other unreactive metals can be mined directly from the Earth's crust, where they occur in their **free elemental state**. Most metals are found combined with other elements in impure ionic compounds, known as **ores**. The metals have to be **extracted** from these ores. Metal oxides, sulfides and carbonates are some of the most important ores.

During extraction from its ore, the metal cations are discharged to form atoms by **gaining** electrons. The extraction of metals is therefore a **reduction process**:

$$M^{n+} + ne^- \longrightarrow M$$

Choosing an extraction method

The extraction method used depends on the **position** of the metal in the reactivity series:

- Metals **high** in the reactivity series (**aluminium and above**) are extracted by **electrolysis of their molten ores**. They require a **powerful method** of reduction because they form very stable ions which are difficult to reduce. Electrolysis is a powerful method, but it uses a lot of energy and is very expensive.
- Metals **lower down** in the series (**zinc and below**) are extracted by **heating their ores with a reducing agent** such as carbon, carbon monoxide or hydrogen. They require a **less powerful method** of reduction than electrolysis because their ions are less stable and easier to reduce. Heating their ores with a reducing agent is a less powerful method, uses less energy and is less expensive than electrolysis.

Extraction of aluminium

Ores: **bauxite** – impure, hydrated aluminium oxide, $Al_2O_3.xH_2O$. Bauxite is the main ore.
cryolite – sodium aluminium fluoride, Na_3AlF_6

The extraction is carried out by **electrolysis**:

- The bauxite is **purified**. This forms pure, anhydrous aluminium oxide, also known as **alumina**, Al_2O_3.
- The **alumina** is dissolved in **molten cryolite** at about 950 °C to separate the ions. The melting point of alumina is 2050 °C and molten alumina is a poor conductor. Dissolving it in molten cryolite reduces its melting temperature which reduces the energy required. The solution produced is also a better conductor than molten alumina.
- The molten solution of alumina in cryolite is **electrolysed** in an electrolytic cell.
  - The **aluminium ions** move towards the **cathode** and are **reduced** to form aluminium atoms:

    $$Al^{3+}(l) + 3e^- \longrightarrow Al(l)$$

    Molten aluminium collects at the bottom of the cell. It is tapped off and made into blocks or sheets.
  - The **oxide ions** move towards the **anode** and are **oxidised** to form oxygen gas:

    $$2O^{2-}(l) \longrightarrow O_2(g) + 4e^-$$

    Oxygen gas is evolved at the anode.

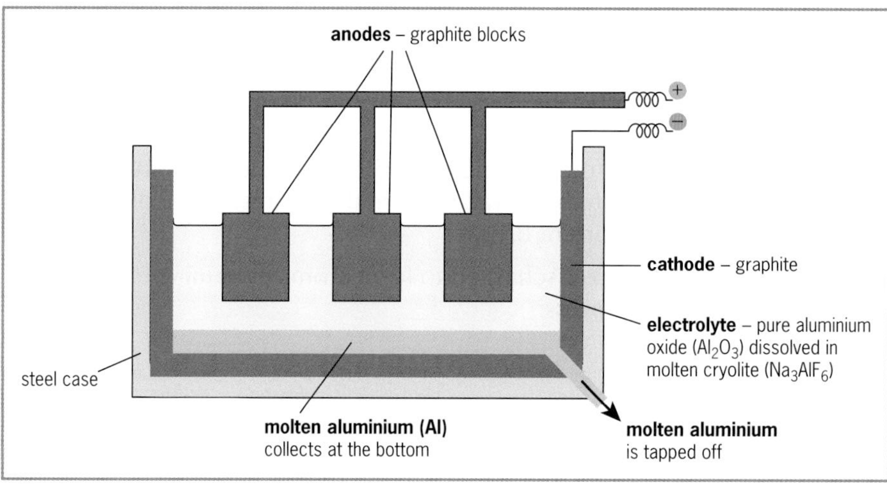

Figure 19.1 *Electrolytic cell for the extraction of aluminium*

Extraction of iron

Ores: **haematite** – impure iron(III) oxide, Fe_2O_3
magnetite – impure iron(II, III) oxide, Fe_3O_4

Extraction of iron from haematite and magnetite is carried out by reducing the ores using the reducing agent, **carbon monoxide (CO)**, in a **blast furnace** (see figure 19.2 p. 161).

- A mixture of the **iron ores**, **coke** (carbon) and **limestone** (calcium carbonate) is added through the top of the furnace.
- **Hot air** is blown in through the bottom of the furnace.
- In the **bottom** part of the furnace, coke reacts with the oxygen in the air to produce **carbon dioxide**:

  $$C(s) + O_2(g) \longrightarrow CO_2(g) \quad \Delta H \text{ –ve}$$

  The reaction is **exothermic**. The heat produced keeps the bottom of the furnace at a temperature of about 1900 °C. The carbon dioxide moves up the furnace.

- As it reaches the **middle** part of the furnace, the carbon dioxide reacts with more of the hot coke, and is reduced to **carbon monoxide**:

$$CO_2(g) + C(s) \longrightarrow 2CO(g)$$

The carbon monoxide moves up the furnace.

- In the **top** part of the furnace, the carbon monoxide **reduces** the iron ores to **iron**:

$$Fe_2O_3(s) + 3CO(g) \longrightarrow 2Fe(l) + 3CO_2(g)$$

and

$$Fe_3O_4(s) + 4CO(g) \longrightarrow 3Fe(l) + 4CO_2(g)$$

The **molten iron** runs to the bottom of the furnace and is tapped off. The iron, known as '**pig iron**', is impure. It contains about 4% carbon, and other impurities such as silicon and phosphorus. Most of the pig iron is purified and converted into an alloy of iron known as **steel**.

The role of the limestone

Iron ores contain a lot of **impurities** which are not removed before the ores are put into the blast furnace. The main impurity is **silicon dioxide** (sand). Limestone is added to remove the silicon dioxide so it does not build up in the furnace.

- In the **top** part of the furnace the heat causes the calcium carbonate to decompose to form **calcium oxide** and carbon dioxide:

$$CaCO_3(s) \longrightarrow CaO(s) + CO_2(g)$$

- Calcium oxide is basic (as it is a metal oxide) and silicon dioxide is acidic (as it is a non-metal oxide). The two then react to form **calcium silicate**, $CaSiO_3$, also known as **slag**:

$$CaO(s) + SiO_2(s) \longrightarrow \underset{\text{slag}}{CaSiO_3(l)}$$

The molten slag runs to the bottom of the furnace where it floats on the molten iron and is tapped off separately. When solidified it can be used as aggregate in concrete for construction purposes, mixed with asphalt and used to build roads, or finely ground and used in the production of cement.

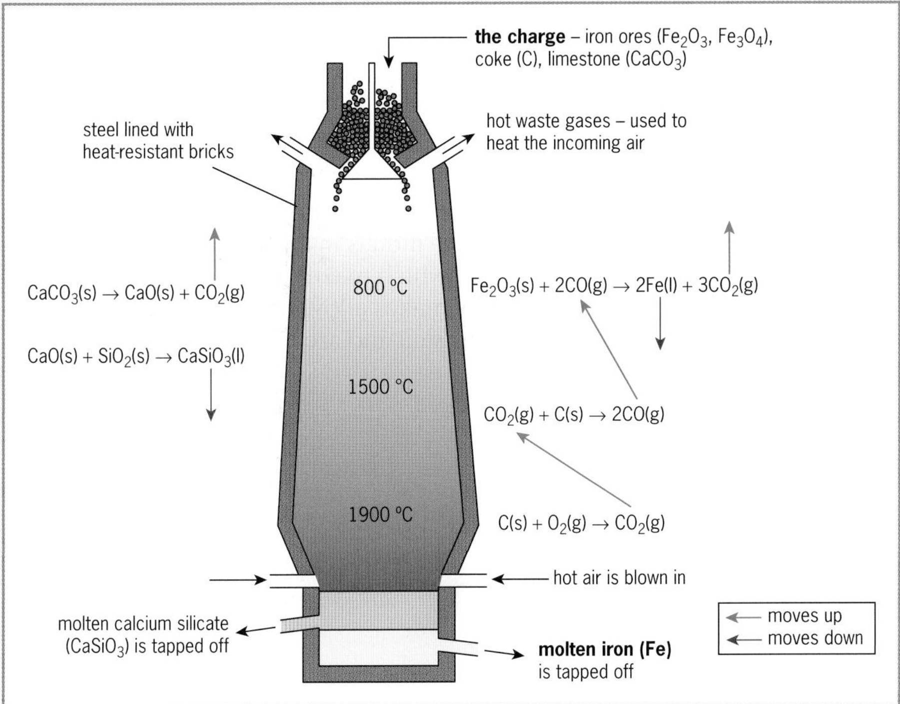

Figure 19.2 *A blast furnace used for extracting iron*

Uses of metals

The **properties** of metals determine their uses.

Table 19.2 *Properties and uses of aluminium, iron and lead*

Metal	Used to make	Properties of the metal
Aluminium	Cans to store drinks.	Resistant to corrosion; malleable so is easily shaped; non-toxic; low density so is light in weight.
	Overhead electrical cables.	Good conductor of electricity; resistant to corrosion; light in weight; ductile so easily drawn out into wires.
	Window frames.	Resistant to corrosion; light in weight.
	Cooking utensils, e.g. saucepans and baking trays.	Good conductor of heat; resistant to corrosion; light in weight; non-toxic; can be polished to have a shiny, attractive appearance.
	Foil used in cooking.	Does not react with the food due to its unreactive aluminium oxide coating; non-toxic; highly reflective so keeps in heat.
Wrought iron (rarely used)	Ornamental iron work.	Malleable and ductile so is easily shaped; strong, which makes it resistant to stress; easily welded.
Lead	Lead-acid batteries, e.g. car batteries.	Good conductor of electricity; very resistant to corrosion.
	Radiation shields, e.g. against X-rays.	High density so prevents radiation from passing through.
	Keels for sailboats; lead weights, e.g. fishing and diving weights.	High density so is heavy; very malleable so is easily shaped.

Note Aluminium, iron and lead are also used extensively to manufacture **alloys** because they are easily alloyed with other metals.

Alloys and uses of alloys

Alloys are **mixtures** of two or more metals, though a few also contain non-metals. Alloys are produced to improve or to modify the properties of metals. The atoms of the metals in an alloy are usually of different sizes. This changes the regular packing of the atoms, which makes it more difficult for them to slide over each other when force is applied. This usually makes alloys **harder** and **stronger** than the pure metals. They are also usually more **resistant to corrosion** and are often used in place of pure metals.

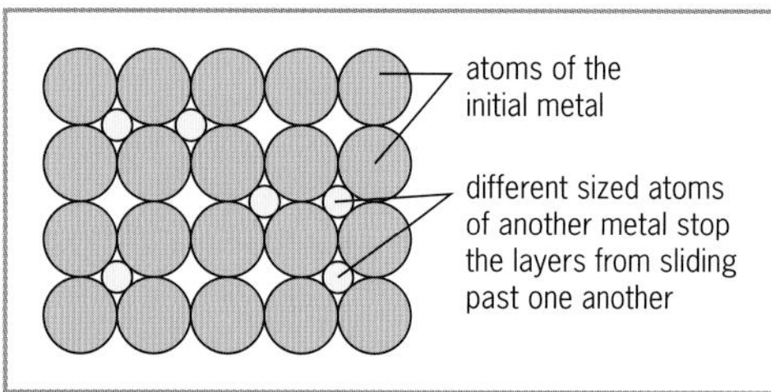

Figure 19.3 *Arrangement of atoms in an alloy*

Table 19.3 *Composition, uses and properties of some common alloys*

Name of alloy	Composition	Uses	Properties of the alloy
Duralumin	Approximately **94% aluminium** alloyed with 4% copper and small amounts of magnesium and manganese.	To construct aircraft and the bodies of motor vehicles.	• Stronger, harder and more workable than aluminium. • Light in weight and resistant to corrosion.
Magnalium	Approximately **95% aluminium** alloyed with 5% magnesium.	To manufacture aircraft and motor vehicle parts.	• Stronger, more workable and lighter in weight than aluminium. • More resistant to corrosion than aluminium.
Mild steel	**Iron** alloyed with less than 0.25% carbon.	To construct buildings, bridges, oil rigs, ships, trains and motor vehicles. To make 'tin cans' to store food; 'tin cans' are made of steel coated in a thin layer of tin which prevents corrosion. To make wire and nails.	• Harder and stronger than iron. • Malleable and ductile so is easily shaped. • Easy to weld.
High carbon steel	**Iron** alloyed with 0.25% to 1.5% carbon.	To make cutting tools, chisels, knives, drill bits and masonry nails.	• Harder than mild steel, but more brittle.
Stainless steel	About **70% iron** alloyed with 20% chromium and 10% nickel.	To make cutlery, kitchen equipment, sinks, catering equipment, surgical equipment and surgical implants. In the construction of modern buildings.	• Harder, stronger and much more resistant to rusting (corrosion) than carbon steels. • Malleable, ductile and easy to work with. • Has a very shiny, attractive appearance.
Cast iron	About **96% iron** and 4% carbon.	To make small castings, e.g. cylinder blocks in engines, railings, gates, manhole covers, hinges and cast iron cookware.	• Easy to cast into exact shapes. • Hard, but more brittle than steel; it shatters rather than bends when hit.
Lead solder	About **60% lead** alloyed with 40% tin.	To join metal items together, e.g. wires and pipes.	• Has a low melting point (lower than lead), so melts easily when joining the metals. • As resistant to corrosion as lead, but harder and stronger.

Revision questions

1 What reactions are used to determine the reactivity series of metals?

2 **a** Place the metals aluminium, copper, sodium, zinc, magnesium, silver and metal **X** in decreasing order of reactivity, given the fact that metal **X** reacts with copper(II) nitrate solution, but does not react with zinc nitrate solution.

b Write an equation for the reaction between **X** and copper(II) nitrate solution, given that **X** forms a divalent cation.

c Explain why **X** reacts with copper(II) nitrate solution but does not react with zinc nitrate solution.

3 Why is the extraction of a metal from its ores a reduction process?

4 Metal **Y** appears just above aluminium in the reactivity series and metal **Z** appears just below zinc. Suggest, with a reason in EACH case, what method should be used to extract EACH metal from its ore.

5 **a** Name the main ore from which aluminium is extracted and give its formula.

b Outline the process by which aluminium is extracted from its ore. Include a relevant equation.

6 **a** By means of THREE equations ONLY, show the reactions involved in extracting iron from the ore haematite (iron(III) oxide) in a blast furnace.

b Limestone (calcium carbonate) is also added to the blast furnace with the ores of iron. Explain, giving appropriate equations, the role of limestone in the extraction process.

7 **a** What is an alloy?

b Why are alloys often used instead of the pure metal?

8 Give TWO reasons for EACH of the following:

a Aluminium is used to make cans to store drinks.

b Lead is used to make lead-acid batteries.

c Stainless steel is used to make cutlery.

d Solder is used to join metal items.

20 Impact of metals on living systems and the environment

Certain **metal ions** play vital roles in living organisms while others can be extremely harmful. At the same time, the **environment** can have a significant effect on metals used in everyday activities.

Corrosion of metals

Corrosion takes place when the surface of a metal is gradually worn away by reacting with chemicals in the environment, mainly **oxygen** and **water vapour**. Certain **pollutants** speed up the process. When a metal corrodes, it is oxidised to form the **metal oxide**. Salts may also form, e.g. the reaction with carbon dioxide forms carbonates. In general, the **higher** a metal is in the reactivity series, the **faster** it corrodes.

The corrosion of aluminium

The corrosion of aluminium is generally **beneficial**. When a fresh piece of aluminium is exposed to air, it immediately forms a layer of **aluminium oxide** (Al_2O_3). This layer **adheres** to the metal surface, does not flake off and is relatively unreactive. It therefore protects the aluminium from further corrosion. The thickness of this layer can be increased by **anodising** (see p. 104).

The corrosion of iron – rusting

The corrosion of iron is **detrimental**. When iron and steel objects are exposed to oxygen and moisture, they immediately begin to corrode forming mainly **hydrated iron(III) oxide, $Fe_2O_3.xH_2O$**, commonly known as **rust**. The corrosion of iron and steel is called **rusting**. Rust does not adhere to the iron below, it **flakes off** instead. This exposes fresh iron to oxygen and moisture, which then rusts and the rust flakes off. The process continues and the iron is gradually worn away.

The importance of metals and their compounds in living organisms

The human body requires certain **metal ions** in relatively large quantities (more than 100 mg per day). These are called **macrominerals** and include calcium, potassium, sodium and magnesium. Others, known as **microminerals** or **trace minerals**, are needed in much smaller quantities, e.g. iron, zinc, manganese, cobalt, copper, molybdenum, selenium and chromium. Plants also require minerals for healthy growth and development.

Table 20.1 *The importance of metals and their compounds in living organisms*

Metal ion	Importance in living organisms
Magnesium	Essential for green plants to produce the green pigment, **chlorophyll**, found in chloroplasts. Chlorophyll absorbs sunlight energy so that plants can manufacture their own food by **photosynthesis**: $6CO_2(g) + 6H_2O(l) \xrightarrow[\text{(absorbed by chlorophyll)}]{\text{sunlight energy}} \underset{\text{glucose}}{C_6H_{12}O_6(aq)} + 6O_2(g)$ A shortage of magnesium causes **chlorosis** where the leaves of plants become yellow. Over 300 biochemical reactions in the human body require magnesium ions because they help many enzymes to function.

Metal ion	Importance in living organisms
Iron	Essential for animals to produce the red pigment, **haemoglobin**, found in red blood cells. Haemoglobin carries oxygen around the body for cells to use in **respiration** to produce energy: $$C_6H_{12}O_6(aq) + 6O_2(g) \longrightarrow 6CO_2(g) + 6H_2O(l) + \text{energy}$$ A shortage of iron leads to **anaemia** where the number of red blood cells is reduced causing tiredness and a lack of energy.
Calcium	Essential to produce **calcium hydroxyapatite** ($Ca_{10}(PO_4)_6(OH)_2$) in the bones and teeth of animals. A shortage of calcium leads to **rickets** in children where the legs become bowed, and **osteoporosis** in adults where the bones become weak and brittle.
Zinc	Important for the functioning of the immune system, for wounds to heal, and for the growth and repair of cells and tissues.
Sodium and **potassium**	Important for impulses to be transmitted along nerves and for muscles to contract.

Note Chlorophyll and haemoglobin are known as **organometallic compounds** because they are organic compounds whose molecules contain metal ions.

Harmful effects of metals and their compounds

The ions of certain transition metals and metalloids, known as **heavy metal ions**, are **toxic** to living organisms, especially when combined with organic compounds to form **organometallic compounds**. These metal ions occur naturally; however, **pollution** caused by human activities is causing their concentrations within the environment to increase.

Pollution *is the contamination of the natural environment by the release of unpleasant and harmful substances into the environment.*

Heavy metal ions are **persistent**, meaning they remain in the environment for a long time. They also become **higher in concentration** (**bioconcentrate**) moving up food chains and can reach harmful levels in top consumers such a birds of prey and large fish. Eating **large fish**, such as sharks, marlin and tuna, is the major source of ingested **mercury** in humans.

Table 20.2 *Harmful effects of metal ions on living organisms*

Metal ion	Sources in the environment	Harmful effects
Lead	Discarded lead-acid batteries; manufacture and recycling of lead-acid batteries; mining of lead ores; extraction and refining of lead from its ores; lead-based paints; car exhaust fumes when using leaded petrol in some countries.	• Damages various body tissues and organs, e.g. the kidneys, liver, bones and nervous system, particularly the brain. • Interferes with the normal formation of red blood cells, which leads to **anaemia**. • Particularly harmful to young children as it reduces IQ and causes behavioural problems and learning disorders.
Arsenic	Mining of certain metals, mainly gold; extraction and refining of metals; burning fossil fuels, especially coal. Volcanic eruptions.	• Causes changes in pigmentation and thickening of the skin, and can cause **cancer**. • Damages the nervous system, heart, lungs and blood vessels.

Metal ion	Sources in the environment	Harmful effects
Cadmium	Discarded nickel-cadmium batteries; cigarette smoke; burning fossil fuels; incinerating waste; extraction and refining of metals.	• Damages the kidneys, liver and respiratory system if inhaled. • Can cause bones to become weakened and fragile, leading to **osteoporosis**.
Mercury	Discarded fluorescent light bulbs; discarded mercury thermometers from laboratories and hospitals; burning coal in coal-fired power plants; extraction and refining of metals.	• Damages the central nervous system, resulting in loss of muscular co-ordination, numbness in the hands and feet, and impaired hearing, sight and speech – a condition known as **Minamata disease**.

A compact fluorescent light bulb containing mercury vapour

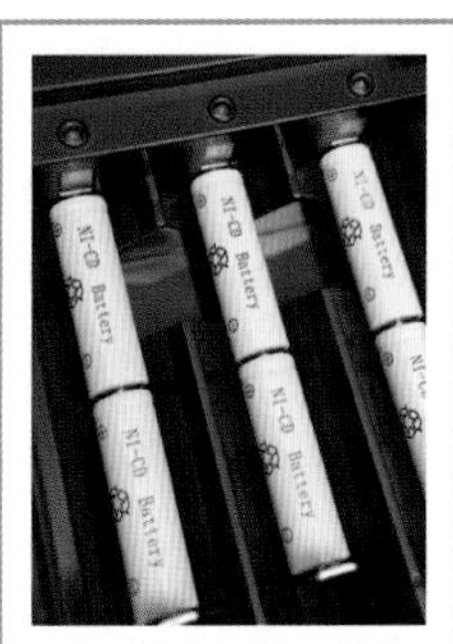

Nickel-cadmium batteries

Disposal of solid waste containing heavy metals

Disposal of solid waste containing **heavy metals** is a serious problem. This waste includes:

- Lead-acid batteries from cars, trucks, other vehicles and boats.
- Nickel-cadmium batteries.
- Fluorescent light bulbs which contain mercury vapour.
- Hospital and laboratory thermometers which contain mercury.

These items should not be disposed of in **landfills** since groundwater and nearby soil could be contaminated, and they should not be **incinerated** because harmful gases containing the metal ions could be released into the air. The more items containing heavy metals that are **recycled**, the more the problem of their disposal will be solved.

Revision questions

1 What happens when metals corrode?

2 What conditions are necessary for iron to rust?

3 Why is the corrosion of aluminium generally beneficial, but the corrosion of iron is detrimental?

4 Explain the importance of the following metals to living organisms:

a magnesium **b** iron **c** calcium **d** zinc

5 Why can the consumption of large fish such as tuna lead to a person developing Minamata disease?

6 Outline the harmful effects of the following heavy metal ions to humans:

a lead **b** arsenic **c** cadmium

21 Non-metals

Non-metals are elements that are found mainly in Groups V, VI, VII and 0 of the periodic table. The atoms of most non-metals have 5, 6, 7 or 8 valence electrons.

Physical properties of non-metals

The physical properties of non-metals vary because of the different ways in which non-metal atoms are bonded in their elements. However, non-metals have the following **general physical properties**:

- Most non-metals have **low** melting and boiling points.
- Non-metals can be **solid**, **liquid** or **gas** at room temperature.
- Non-metals are **poor** conductors of electricity and heat (except graphite).
- Non-metals in the solid state are **weak** and **brittle**.
- Non-metals in the solid state are **dull** in appearance.
- Non-metals usually have **low** densities.

Table 21.1 *Specific properties of certain non-metals*

Non-metal	Properties
Hydrogen (H_2)	A gas at room temperature. Colourless, odourless and tasteless. Virtually insoluble in water. Less dense than air; hydrogen is the lightest known element.
Chlorine (Cl_2)	A yellow-green, poisonous gas with a strong odour. Moderately soluble in water. Denser than air.
Oxygen (O_2)	A gas at room temperature. Colourless, odourless and tasteless. Slightly soluble in water. Slightly denser than air.
Carbon (C)	Has two main allotropes, diamond and graphite (see p. 41): • **Diamond**: ◆ An extremely hard, transparent, colourless, sparkling solid. ◆ Has a very high melting point. ◆ Does not conduct electricity. • **Graphite**: ◆ A soft, flaky, opaque, dark grey solid. ◆ Has a very high melting point. ◆ Conducts electricity.
Sulfur (S)	A solid at room temperature. Yellow in colour.
Nitrogen (N_2)	A gas at room temperature. Colourless, odourless and tasteless. Virtually insoluble in water. Slightly less dense than air.

Chemical properties of non-metals

- When they react with **metals**, non-metals ionise by **gaining** electrons to form negative **anions**. The non-metal behaves as an **oxidising agent** since it removes electrons from the metal (it causes the metal to **lose** electrons). This results in the formation of **ionic compounds**.

  $N + ne^- \longrightarrow N^{n-}$

- When non-metals react with other **non-metals**, they **share** valence electrons. This results in the formation of **covalent compounds**. The non-metals may behave as **oxidising** or **reducing agents**.

Reactions of non-metals with metals and oxygen

- Reactions between non-metals and **metals** are **redox reactions** in which the non-metal acts as the oxidising agent and the metal acts as the reducing agent.
- Reactions between non-metals and **oxygen** are **redox reactions** in which oxygen acts as the oxidising agent and the other non-metal acts as the reducing agent. The product of the reaction is a **non-metal oxide**. Most non-metal oxides are **acidic**, e.g. carbon dioxide (CO_2), sulfur dioxide (SO_2), sulfur trioxide (SO_3) and nitrogen dioxide (NO_2). A few are **neutral**, e.g. water (H_2O), carbon monoxide (CO) and nitrogen monoxide (NO).

Table 21.2 *Reactions of some non-metals with metals and with oxygen*

Non-metal	Reaction with metals (using magnesium as an example)	Reaction with oxygen
Hydrogen	Produces **metal hydrides**: $Mg(s) + H_2(g) \longrightarrow MgH_2(s)$ magnesium hydride	Burns with a very pale blue flame to produce **water** as steam: $2H_2(g) + O_2(g) \longrightarrow 2H_2O(g)$
Chlorine	Produces **metal chlorides**: $Mg(s) + Cl_2(g) \longrightarrow MgCl_2(s)$ magnesium chloride	—
Oxygen	Produces **metal oxides**: $2Mg(s) + O_2(g) \longrightarrow 2MgO(s)$ magnesium oxide	—
Carbon	—	Burns to produce either carbon monoxide or carbon dioxide: • In a limited supply of oxygen, **carbon monoxide** is formed: $2C(s) + O_2(g) \longrightarrow 2CO(g)$ • In a plentiful supply of oxygen, **carbon dioxide** is formed: $C(s) + O_2(g) \longrightarrow CO_2(g)$
Sulfur	Produces **metal sulfides**: $Mg(s) + S(s) \longrightarrow MgS(s)$ magnesium sulfide	Burns with a blue flame to form **sulfur dioxide**: $S(s) + O_2(g) \longrightarrow SO_2(g)$
Nitrogen	Produces **metal nitrides**: $3Mg(s) + N_2(g) \longrightarrow Mg_3N_2(s)$ magnesium nitride	Reacts with oxygen, if the temperature is high enough, to form **nitrogen monoxide**: $N_2(g) + O_2(g) \longrightarrow 2NO(g)$

Behaviour of non-metals as oxidising and reducing agents

Unless reacting with oxygen, non-metals usually act as **oxidising agents**. However, hydrogen, carbon and sulfur can also act as reducing agents.

Non-metals acting as oxidising agents

- **All** non-metals act as oxidising agents when reacting with **metals** (see Table 21.2, p. 169).
- **Oxygen** and **chlorine** act as oxidising agents in **all** reactions.

e.g.

$$2H_2(g) + O_2(g) \longrightarrow 2H_2O(g)$$
$$H_2(g) + Cl_2(g) \longrightarrow 2HCl(g)$$
$$CH_4(g) + 2O_2(g) \longrightarrow CO_2(g) + 2H_2O(g)$$
$$2KI(aq) + Cl_2(g) \longrightarrow 2KCl(aq) + I_2(aq)$$

Non-metals acting as reducing agents

- **Hydrogen**, **carbon, sulfur** and **nitrogen** act as reducing agents when reacting with **oxygen** (see Table 21.2, p. 169).
- **Hydrogen** and **carbon** act as reducing agents when reacting with **metal oxides**. They reduce the metal ions to produce metal atoms.

e.g.

$$PbO(s) + H_2(g) \longrightarrow Pb(s) + H_2O(l)$$
$$2CuO(s) + C(s) \longrightarrow 2Cu(s) + CO_2(g)$$

Laboratory preparation of gases

When deciding on the **method** to prepare a gas in the laboratory, certain properties of the gas must be considered.

Table 21.3 *Properties to be considered when choosing a method to prepare a gas*

Property to consider	Reason
Solubility of the gas in water.	To determine if the gas can be collected by bubbling it through water if it does not need to be dry.
Reactivity of the gas with different drying agents.	To determine which drying agents can be used if a dry gas is required.
Density of the gas compared to the density of air.	To help choose the correct method to collect the gas if it has been dried.

Laboratory preparation of carbon dioxide and oxygen

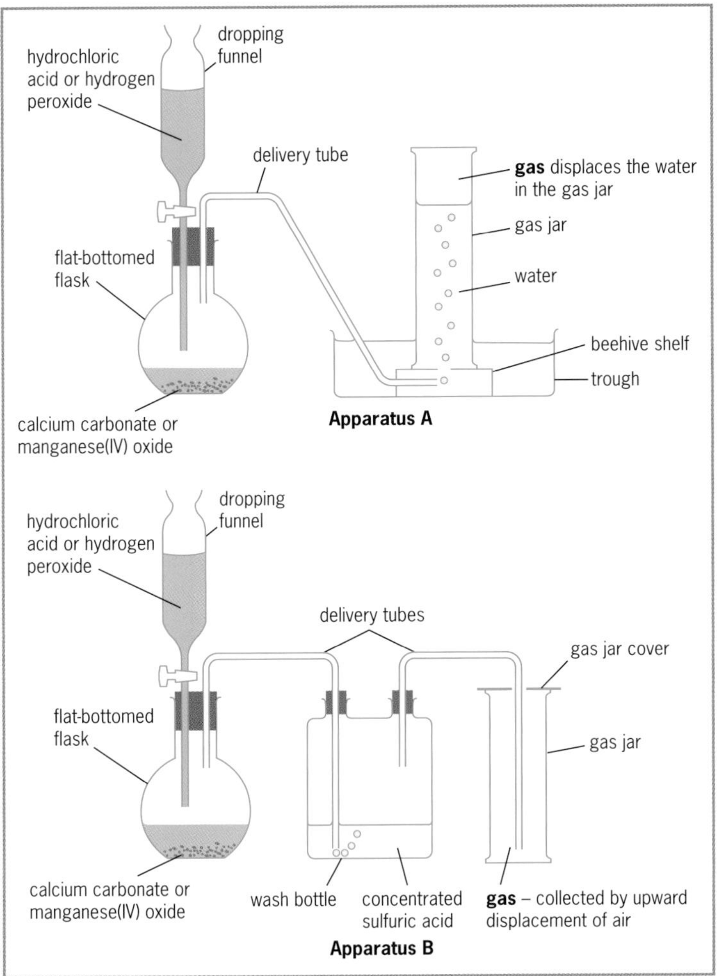

Figure 21.1 *Apparatus used for the laboratory preparation of oxygen and carbon dioxide*

Table 21.4 *Laboratory preparation of carbon dioxide and oxygen*

	Carbon dioxide (CO_2)	**Oxygen (O_2)**
Method	Reacting a **carbonate** with an **acid**, usually calcium carbonate and hydrochloric acid: $CaCO_3(s) + 2HCl(aq) \longrightarrow CaCl_2(aq) + CO_2(g) + H_2O(l)$	Decomposition of **hydrogen peroxide** using a **catalyst** of manganese(IV) oxide: $2H_2O_2(l) \xrightarrow{MnO_2} 2H_2O(l) + O_2(g)$
Apparatus	For **wet** carbon dioxide: apparatus **A** For **dry** carbon dioxide: apparatus **B**	For **wet** oxygen: apparatus **A** For **dry** oxygen: apparatus **B**
Drying agent	• **Concentrated sulfuric acid** in a wash bottle, or • Anhydrous **calcium chloride** in a U-tube (as shown in Figure 21.2, p. 172).	• **Concentrated sulfuric acid** in a wash bottle, or • Anhydrous **calcium chloride** in a U-tube, or • **Calcium oxide** in a U-tube.
Collection of the dry gas	**Upward displacement of air** since it is denser than air.	**Upward displacement of air** since it is slightly denser than air.

Note Calcium carbonate and sulfuric acid cannot be used to prepare carbon dioxide. The reaction quickly stops because it makes **insoluble** calcium sulfate which forms a layer around the calcium carbonate crystals and prevents them from continuing to react.

Laboratory preparation of ammonia

Ammonia can be prepared by heating a **base** with an **ammonium salt**, e.g. calcium hydroxide and ammonium chloride:

$$Ca(OH)_2(s) + 2NH_4Cl(s) \xrightarrow{\text{heat}} CaCl_2(s) + 2NH_3(g) + 2H_2O(g)$$

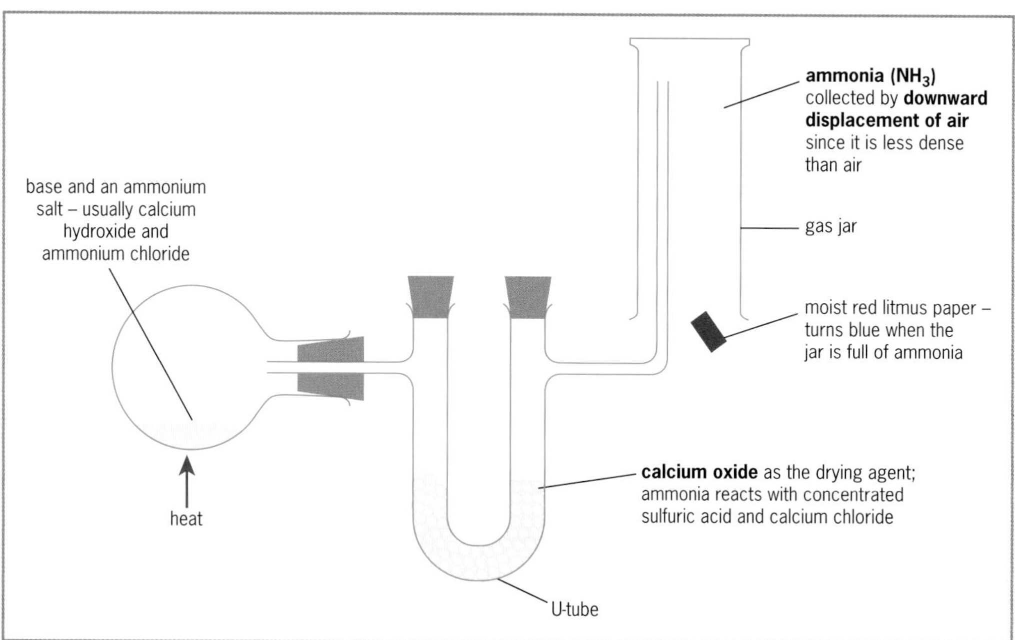

Figure 21.2 *The laboratory preparation of dry ammonia*

Note Ammonia reacts with water to form ammonium hydroxide solution, therefore **cannot** be collected by bubbling it through water.

Uses of carbon dioxide and oxygen

Because of their **properties**, carbon dioxide and oxygen have many uses.

Uses of carbon dioxide

Table 21.5 *Uses and properties of carbon dioxide*

Uses	Properties
In **fire extinguishers** when liquefied under pressure.	Non-flammable. Denser than air so it smothers the flames and keeps oxygen out.
To make **carbonated soft drinks**.	Dissolves in the drink under pressure and bubbles out when pressure is released. Adds a pleasant tingle and taste.
As a **refrigerant** in the solid state ('dry ice').	Sublimes at –78.5 °C so keeps frozen foods at a very low temperature and does not leave a liquid residue when it sublimes to a gas.
As an **aerosol propellant** for certain foods, e.g. whipped cream.	Relatively inert.

Uses of oxygen

Oxygen is essential for living organisms to carry out **aerobic respiration** and produce **energy**. It is therefore used:

- In **hospitals** to help patients with breathing difficulties and to ease certain medical disorders, including emphysema, asthma, chronic bronchitis and heart disease.
- On **aeroplanes** and **submarines** for people onboard to breathe in emergencies.
- In **spacesuits** for astronauts to breathe.

Oxygen is essential for **combustion** to occur. It is therefore used:

- In **welding torches** to burn the acetylene or hydrogen and produce extremely high temperature flames.
- In the liquid state it is used to burn the fuel to produce the thrust in **rocket engines**.

Uses of non-metals and their compounds

Non-metals and their compounds have a great many uses.

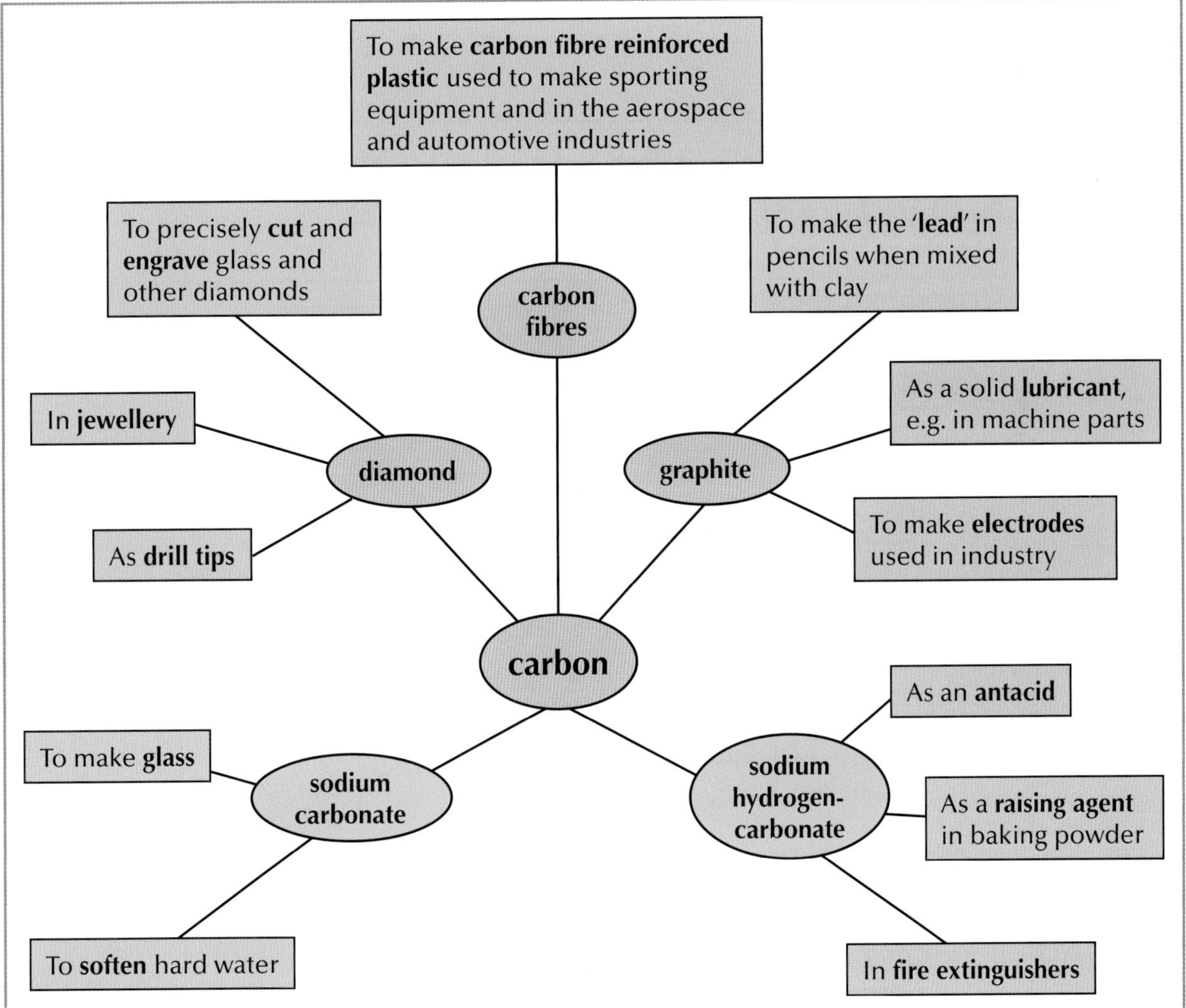

Figure 21.3 *The uses of carbon and its compounds*

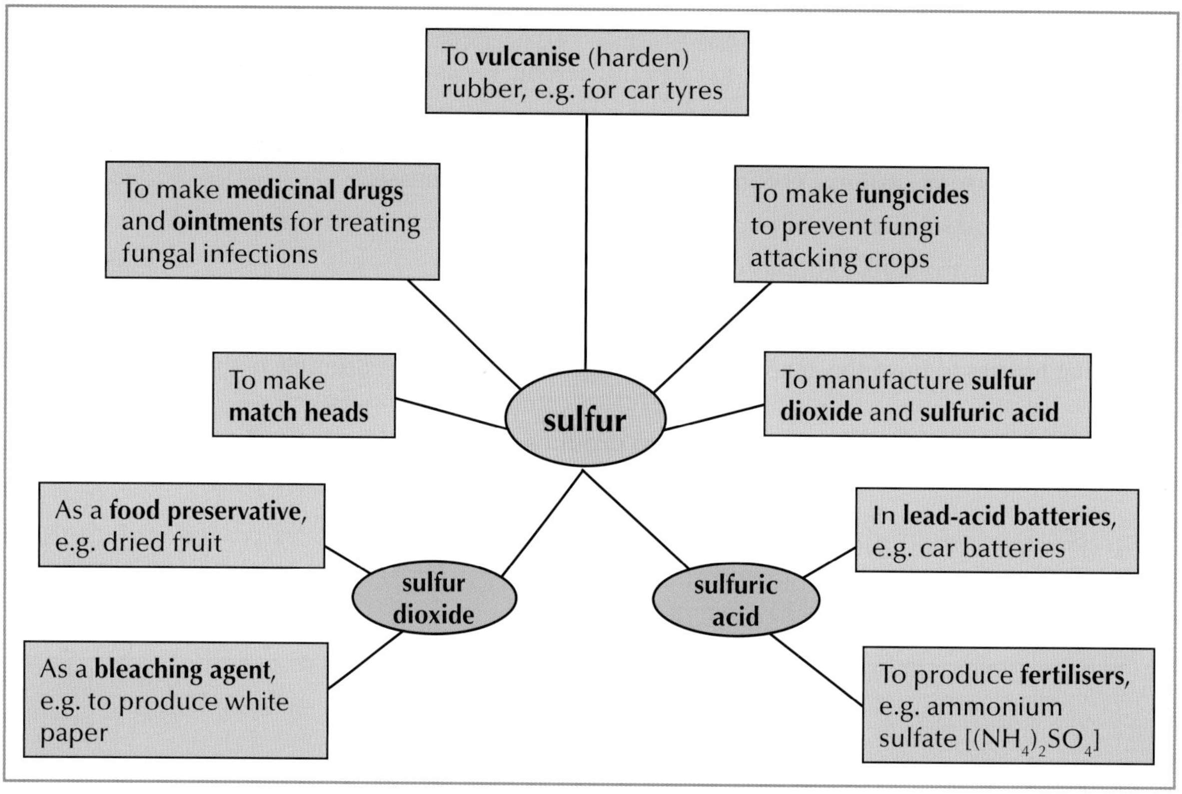

To vulcanise (harden) rubber, e.g. for car tyres
To make medicinal drugs and ointments for treating fungal infections
To make fungicides to prevent fungi attacking crops
To make match heads
sulfur
To manufacture sulfur dioxide and sulfuric acid
As a food preservative, e.g. dried fruit
In lead-acid batteries, e.g. car batteries
sulfur dioxide
sulfuric acid
As a bleaching agent, e.g. to produce white paper
To produce fertilisers, e.g. ammonium sulfate [$(NH_4)_2SO_4$]

Figure 21.4 *The uses of sulfur and its compounds*

To make the striking surface on the side of safety match boxes
To produce fireworks, flares and explosives
To produce pesticides
phosphorus
To manufacture phosphoric acid
To make match heads
To remove rust
phosphoric acid
phosphorus sulfides, e.g. P_4S_3
To produce fertilisers, e.g. ammonium phosphate [$(NH_4)_3PO_4$]

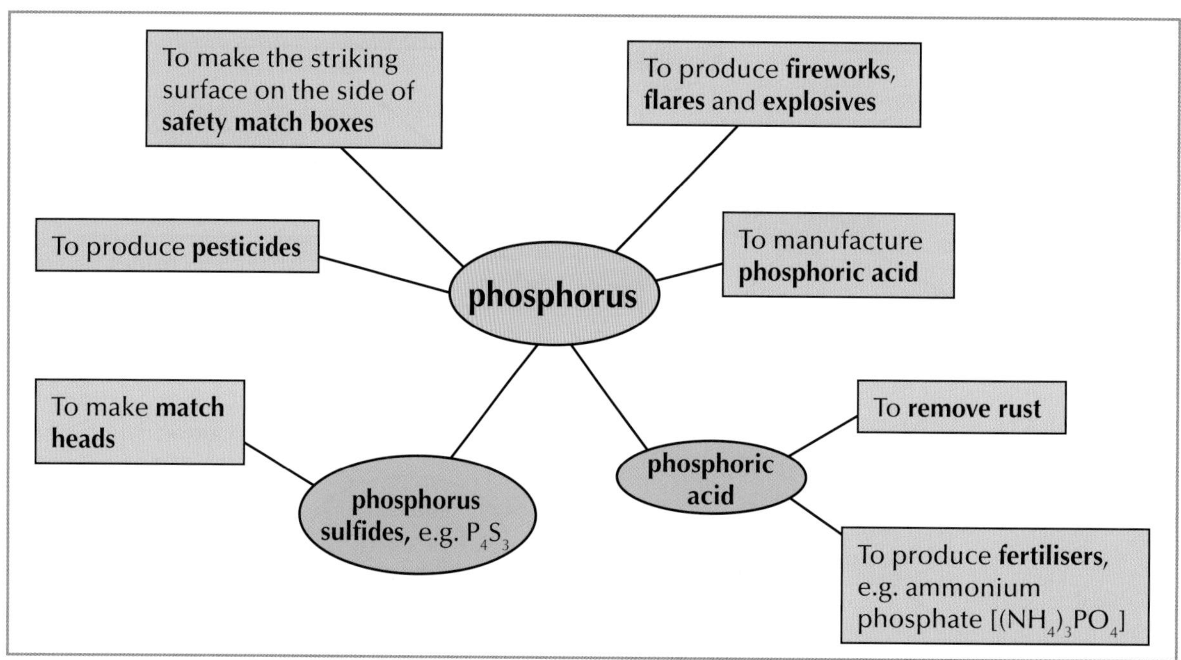

Figure 21.5 *The uses of phosphorus and its compounds*

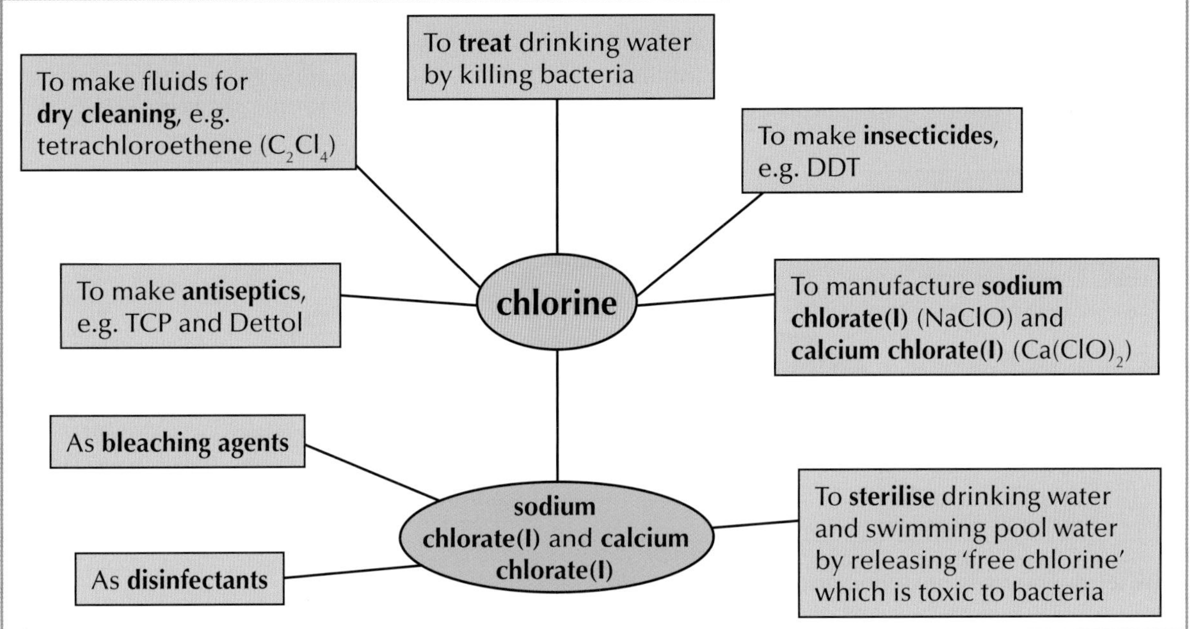

Figure 21.6 *The uses of chlorine and its compounds*

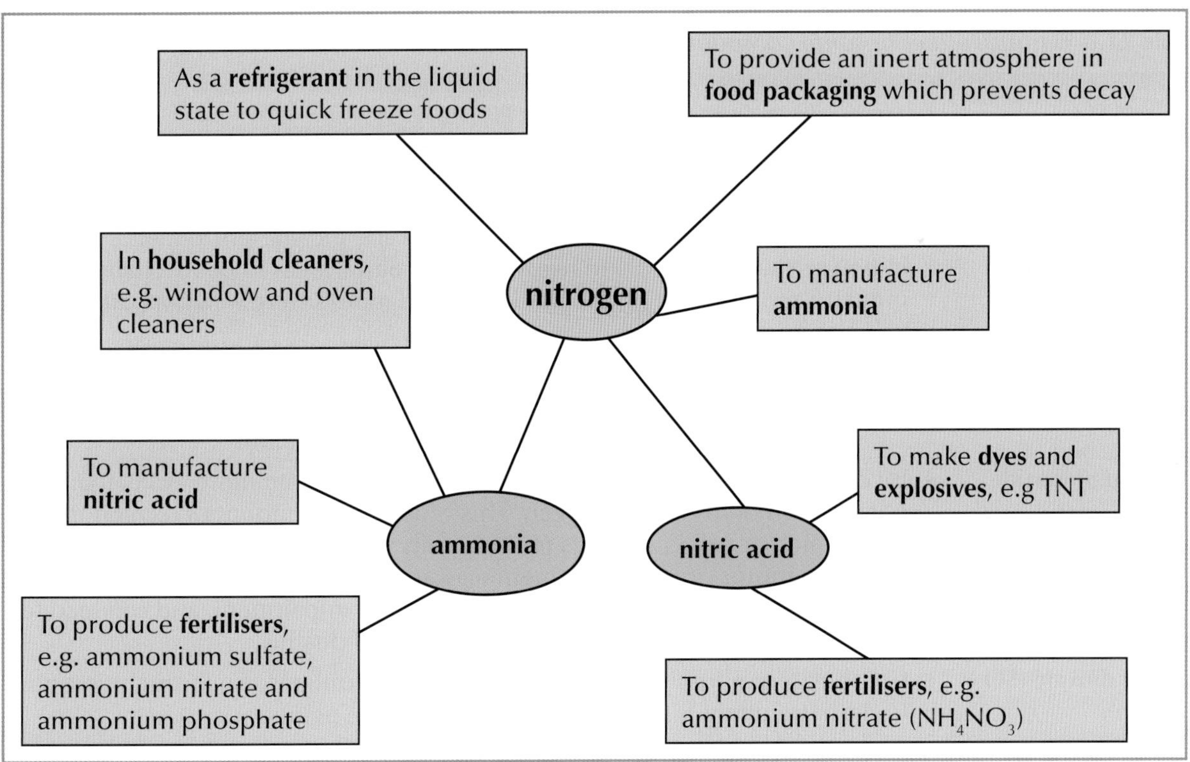

Figure 21.7 *The uses of nitrogen and its compounds*

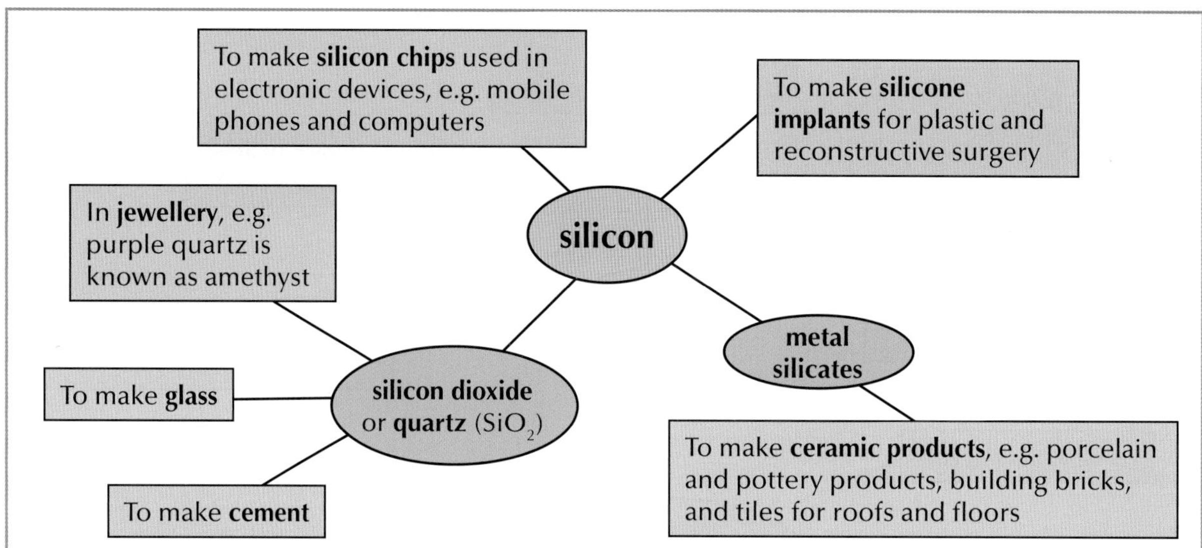

Figure 21.8 *The uses of silicon and its compounds*

Harmful effects of non-metals and their compounds

Some compounds of non-metals can be harmful to living organisms and the environment. Human activities are causing an increase in their concentrations within the environment.

Table 21.6 *Harmful effect of some compounds of non-metals*

Pollutant	Sources in the environment	Harmful effects on living organisms and the environment
Carbon dioxide (CO_2)	• Complete combustion of fossil fuels, particularly in power stations, industry, motor vehicles and aeroplanes.	• Builds up in the upper atmosphere enhancing the **greenhouse effect** and **global warming**, which is causing polar ice caps and glaciers to melt, sea levels to rise, low lying coastal areas to flood, changes in global climate and more severe weather patterns. • Some is absorbed by oceans causing **ocean acidification**, which is expected to affect the ability of shellfish to produce their shells, and of reef building corals to produce their skeletons.
Carbon monoxide (CO)	• Incomplete combustion of fossil fuels which occurs mainly in motor vehicles.	• Combines with haemoglobin more easily than oxygen. This reduces the amount of oxygen that gets to body cells which reduces respiration and mental awareness. It causes dizziness, headaches and visual impairment, and can lead to unconsciousness and death.
Sulfur dioxide (SO_2)	• Combustion of fossil fuels, particularly coal and heavy oils in power stations and industry.	• Causes respiratory problems, e.g. bronchitis, and reduces the growth of plants. • Dissolves in rainwater, forming **acid rain**. Acid rain decreases the pH of soil, damages plants, harms animals, corrodes buildings, and causes lakes, streams and rivers to become acidic and unsuitable for aquatic organisms. • Combines with water vapour and smoke, forming **smog** which causes respiratory problems, e.g. bronchitis, asthma and lung disease.

Pollutant	Sources in the environment	Harmful effects on living organisms and the environment
Oxides of nitrogen (NO and NO_2)	• Combustion at high temperatures in power stations and engines of motor vehicles which causes nitrogen and oxygen in the air to react.	• Very toxic. Cause lung damage and even at low concentrations they irritate the respiratory system, skin and eyes. • Cause leaves to die and reduce plant growth. • Dissolve in rainwater producing **acid rain** (see previous page).
Hydrogen sulfide (H_2S)	• Decomposing organic waste in farmyards, landfills and garbage dumps. • A waste product from petroleum refineries.	• Extremely toxic. Even low concentrations irritate the eyes and respiratory system. • Combines readily with haemoglobin in the same way as carbon monoxide (see previous page).
Carbon particles (C)	• Combustion of fossil fuels. • Bush fires and cigarette smoke.	• Coat leaves which reduces photosynthesis, and blacken buildings. • Combine with water vapour and sulfur dioxide to form **smog** (see above).
Chlorofluorocarbons (CFCs)	• Used in refrigerators and air conditioners as a refrigerant, and in some aerosol sprays as a propellant.	• Break down the **ozone layer** in the upper atmosphere. This allows more ultraviolet light to reach the Earth's surface which is leading to more people developing skin cancer, cataracts and depressed immune systems.
Pesticides, e.g. **insecticides, fungicides and herbicides**	• Used in agriculture to control pests, diseases and weeds. • Used to control vectors of disease, e.g. mosquitoes.	• Become **higher in concentration** up food chains and can harm top consumers. • Can harm **useful** organisms as well as harmful ones, e.g. bees which are crucial for pollination in plants.
Nitrate ions (NO_3^-) and **phosphate ions (PO_4^{3-})**	• Chemical fertilisers used in agriculture. • Synthetic detergents.	• Cause **eutrophication**, i.e. the rapid growth of green plants and algae in lakes, ponds and rivers, which causes the water to turn green. The plants and algae begin to die and are decomposed by aerobic bacteria which multiply and use up the dissolved oxygen. This causes other aquatic organisms to die, e.g. fish.

Disposal of solid waste containing plastics

Plastics are organic compounds composed of non-metallic elements. They are made mainly from hydrocarbons obtained from natural gas, crude oil and coal. Most plastics are **non-biodegradable**, so they remain in the environment for a very long time. Because of this, getting rid of solid waste containing plastics is a major problem:

- **Toxic gases** are produced when plastics burn. Disposal of plastics in incinerators can cause air pollution and health problems.
- **Toxic chemicals** are continually released from some plastics. These chemicals can contaminate soil and groundwater when plastics are disposed of in landfills.
- About 25% of all solid waste going to **landfills** is composed of plastics. More and more land is being used up to dispose of these plastics.
- Plastics often end up in **lakes, rivers** and **oceans** when not disposed of correctly and can harm aquatic organisms (see p. 150).

The more plastic items that are **recycled**, the more the problem of their disposal will be solved.

Solid waste containing plastics

Revision questions

1 Give FOUR physical properties that are typical of most non-metals.

2 State whether non-metals behave as oxidising or reducing agents when they react with metals. Give a reason for your answer.

3 Explain how hydrogen can behave as both an oxidising agent and a reducing agent by referring to TWO different reactions of hydrogen. Your answer must include relevant equations.

4
- **a** Draw a labelled diagram of the apparatus you would use to prepare and collect dry carbon dioxide in the laboratory. Your labels must identify the reactants and the drying agent you would use.
- **b** Name an alternative drying agent that could be used.
- **c** Explain the reason for collecting the gas using the method shown in your diagram.

5 Give THREE uses of carbon dioxide and state ONE reason why the gas is suitable for EACH use.

6
- **a** Name the non-metal which is used to vulcanise rubber and state what happens when rubber is vulcanised.
- **b** Name THREE different non-metals whose compounds are used to manufacture fertilsers and give a named example of the fertiliser in EACH case.

7 Give TWO uses of EACH of the following non-metals:

a carbon **b** chlorine **c** nitrogen **d** silicon

8 Discuss how EACH of the following is harmful to human health and the environment when released into the environment by man's activities:

a sulfur dioxide **b** carbon dioxide **c** chlorofluorocarbons **d** nitrate ions

22 Water

Water molecules are **polar** (see p. 37). The partial positively charged hydrogen atoms and partial negatively charged oxygen atoms of water molecules are attracted to each other forming **hydrogen bonds**. Hydrogen bonds are generally **stronger** than the intermolecular forces between other molecules and they have a significant effect on the **physical properties** of water. These properties are important to **living organisms**.

The unique properties of water

- **Water has a maximum density at 4 °C**

  When water is cooled to 4 °C it **contracts** and becomes denser. If it is cooled below 4 °C, it starts to **expand** and become less dense. It continues to expand until it freezes at 0 °C. As a result, ice at 0 °C **floats** on water as it forms.

  When a pond, lake or river freezes in winter, ice forms on the surface and the warmer, denser water remains below. **Aquatic organisms** are able to survive in the water beneath the ice.

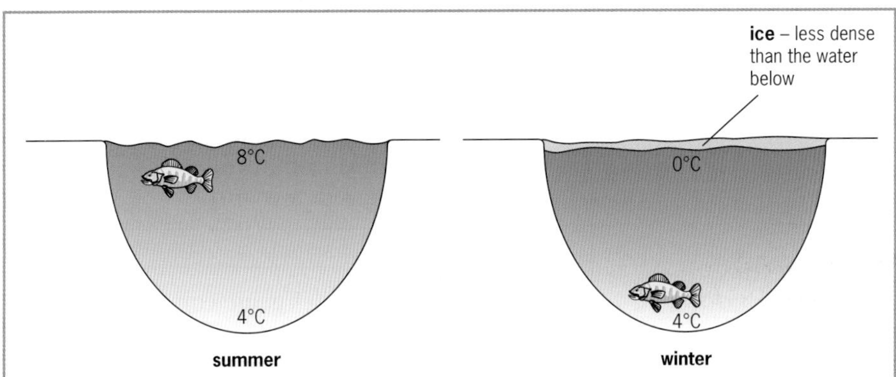

Figure 22.1 *Water temperatures in a pond or lake in the summer and winter*

- **Water has a high heat of vaporisation**

  A lot of heat energy is required to change liquid water to water vapour.

  When water evaporates from the surface of an organism it removes a lot of heat energy from the organism, making sweating and transpiration efficient **cooling** mechanisms.

- **Water has relatively high melting and boiling points**

  Compared with other molecules of a similar size, water has relatively high melting and boiling points (0 °C and 100 °C, respectively), therefore water is **not very volatile**.

  At the temperatures experienced on Earth, most water is in the liquid state. This means that seas, rivers, lakes and ponds exist and provide an environment for aquatic organisms.

- **Water has a high specific heat capacity**

  It requires a **lot of heat energy** to increase the temperature of water by 1 °C. As a result, water can absorb a lot of heat energy without its temperature changing very much.

  - Since the bodies of living organisms contain between 60% and 70% water, they can absorb a lot of heat energy without their body temperatures changing very much, so can survive in extremes of temperature.
  - As atmospheric temperatures change, the temperatures of large bodies of water, such as lakes and seas, do not change very much. Therefore, aquatic organisms do not experience extreme fluctuations in the temperature of their environment.

- **Water dissolves a large number of substances**

  Being **polar**, water can dissolve both ionic and polar covalent compounds. This is **important** to living organisms.

  - Water dissolves chemicals in the cells of organisms so that **chemical reactions** such as respiration can occur.
  - Water dissolves useful substances, such as food and mineral salts, so they can be **absorbed** into the bodies of organisms and **transported** around their bodies.
  - Water dissolves waste and harmful substances so they can be **excreted** from living organisms, e.g. urea dissolves in water forming urine.

  The solvent properties of water can also be **detrimental** because they can cause water to become **hard** or **polluted** and mineral salts to be **leached** out of the soil (see below).

Consequences of water's solvent properties

Water pollution

Water becomes **polluted** when it dissolves harmful substances in the environment.

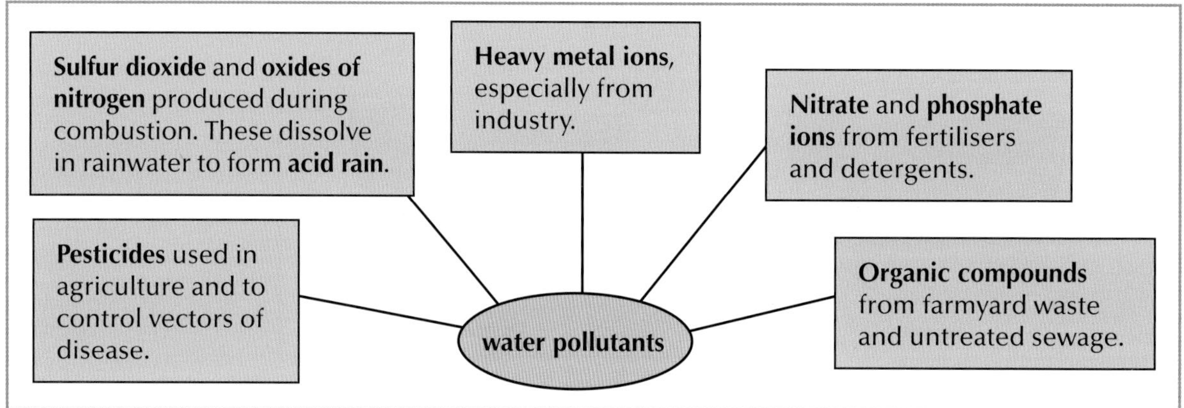

Figure 22.2 *The main water pollutants*

Leaching

Leaching occurs as water passes through the soil and dissolves water-soluble substances. This makes the soil **less fertile** since it can take mineral salts out of the reach of plant roots.

Water hardness

Dissolved **calcium** and **magnesium salts** cause water to become hard. **Hard water** does not lather easily with soap, whereas **soft water** lathers easily with soap. When soap, i.e. sodium octadecanoate ($C_{17}H_{35}COONa$), is added to water containing dissolved Ca^{2+} or Mg^{2+} ions (hard water), insoluble calcium and magnesium octadecanoate form. This is commonly called **scum**:

e.g. $$\underset{\text{soap}}{2C_{17}H_{35}COONa(aq)} + Ca^{2+}(aq) \longrightarrow \underset{\text{scum}}{(C_{17}H_{35}COO)_2Ca(s)} + 2Na^{+}(aq)$$

Hard water is **inconvenient** because:

- It causes unpleasant scum to form. Scum discolours clothes and forms a grey, greasy layer around sinks, baths and showers.
- It **wastes** soap. Soap only lathers when all the Ca^{2+} or Mg^{2+} ions have been precipitated out as scum.
- It causes **lime scale** (see p. 181) to be deposited in kettles, boilers and hot water pipes. This wastes electricity and can block pipes.

There are **two** types of water hardness:

- **Temporary hardness**

  Dissolved **calcium hydrogencarbonate** ($Ca(HCO_3)_2$) and **magnesium hydrogencarbonate** ($Mg(HCO_3)_2$) cause temporary hardness. It is found in limestone-rich areas. When rainwater containing dissolved carbon dioxide passes through limestone (calcium carbonate, $CaCO_3$), soluble calcium hydrogencarbonate forms which dissolves in the rainwater:

  $$CaCO_3(s) + H_2O(l) + CO_2(g) \longrightarrow Ca(HCO_3)_2(aq)$$

  Temporary hardness **can** be removed by **boiling** the water (see below).

- **Permanent hardness**

  Dissolved **calcium sulfate** ($CaSO_4$) and **magnesium sulfate** ($MgSO_4$) cause permanent hardness. Permanent hardness **cannot** be removed by **boiling.**

Treatment of water for domestic use

To make water safe for domestic use, it is **treated** using various different methods.

Large-scale treatment of water

The following steps can be used to treat water before it is piped to homes:

- **Flocculation.** Certain chemicals, such as alum, are added so that fine suspended solid particles clump together to form larger particles called **floc**.
- **Sedimentation.** The floc is allowed to settle.
- **Filtration.** The clear water above the floc is passed through **filters** to remove any remaining particles. This also removes some microorganisms, e.g. bacteria and viruses.
- **Chlorination. Chlorine gas** or **monochloroamine** (NH_2Cl) are added to kill any remaining bacteria and viruses.

Methods to treat water in the home

- **Boiling**

  Microorganisms can be killed by **boiling** water for 15 minutes.

- **Filtering**

  Fibre filters can be used to remove suspended particles. **Carbon filters** containing activated charcoal can be used to remove dissolved organic compounds, odours and unpleasant tastes.

- **Chlorinating**

  Microorganisms can be destroyed by adding **sodium chlorate(I)** solution ($NaClO$) or **calcium chlorate(I)** tablets ($Ca(ClO)_2$) to water.

Methods to soften hard water

Hard water can be converted to **soft** water by removing the dissolved Ca^{2+} or Mg^{2+} ions. All methods remove both temporary and permanent hardness, except boiling.

- **Boiling**

  Boiling removes **temporary hardness** only by causing dissolved calcium hydrogencarbonate and magnesium hydrogencarbonate to decompose:

  e.g. $$Ca(HCO_3)_2(aq) \xrightarrow{\text{heat}} CaCO_3(s) + H_2O(l) + CO_2(g)$$

  The insoluble carbonates precipitate out thereby removing the Ca^{2+} and Mg^{2+} ions. The calcium carbonate produced is also known as **lime scale** or **kettle fur**.

- **Adding washing soda, i.e. sodium carbonate**

  Sodium carbonate, added to hard water, causes dissolved Ca^{2+} or Mg^{2+} ions to precipitate out as **insoluble** calcium carbonate and magnesium carbonate:

  Removing **temporary hardness**:

  e.g. $Ca(HCO_3)_2(aq) + Na_2CO_3(aq) \longrightarrow CaCO_3(s) + 2NaHCO_3(aq)$

  Removing **permanent hardness**:

  e.g. $CaSO_4(aq) + Na_2CO_3(aq) \longrightarrow CaCO_3(s) + Na_2SO_4(aq)$

  Ionically: $Ca^{2+}(aq) + CO_3^{2-}(aq) \longrightarrow CaCO_3(s)$

- **Ion-exchange**

  Water is slowly passed through an ion-exchange column in a water-softening device. The column contains an **ion-exchange resin** called **zeolite, Na_2Z**. Any Ca^{2+} and Mg^{2+} ions in the water **displace** the Na^+ ions and are absorbed into the zeolite. The Na^+ ions enter the water but do not cause it to be hard:

  e.g. $Ca^{2+}(aq) + \underset{\text{ion-exchange resin}}{Na_2Z(s)} \longrightarrow CaZ(s) + 2Na^+(aq)$

- **Distillation**

  Water is boiled and the steam is condensed to form pure **distilled water** which is collected in a separate container. Any dissolved salts and microorganisms are left behind.

- **Reverse osmosis**

  Water is forced through a **differentially permeable** membrane under pressure and dissolved substances remain behind on the pressurised side. This removes Ca^{2+} and Mg^{2+} ions in the water as well as other ions, contaminants and most microorganisms.

Revision questions

1. Why does water have unique properties?

2. Explain the reason for EACH of the following:

   a Ice floats on the surface of ponds in winter.

   b Sweating is an efficient cooling mechanism.

3. Give TWO reasons why the high specific heat capacity of water is important to living organisms.

4. **a** Give THREE advantages of water's good solvent properties.

   b Give THREE disadvantages of water's good solvent properties.

5. **a** What is 'hard water'?

   b What causes water to become 'hard'?

6. Explain how EACH of the following can soften hard water and support EACH answer with a relevant equation:

   a boiling the water

   b adding sodium carbonate

   c using an ion-exchange resin

7. Identify THREE ways water can be treated in the home other than softening.

23 Green Chemistry

Green Chemistry, also known as **sustainable chemistry**, provides a framework for chemists to use when designing new materials, chemical products and processes. Its primary aim is to reduce the flow of **chemical pollutants** into the environment.

Green Chemistry is the utilisation of a set of principles in the design, manufacture and application of chemical products that reduces or eliminates the use and generation of hazardous substances.

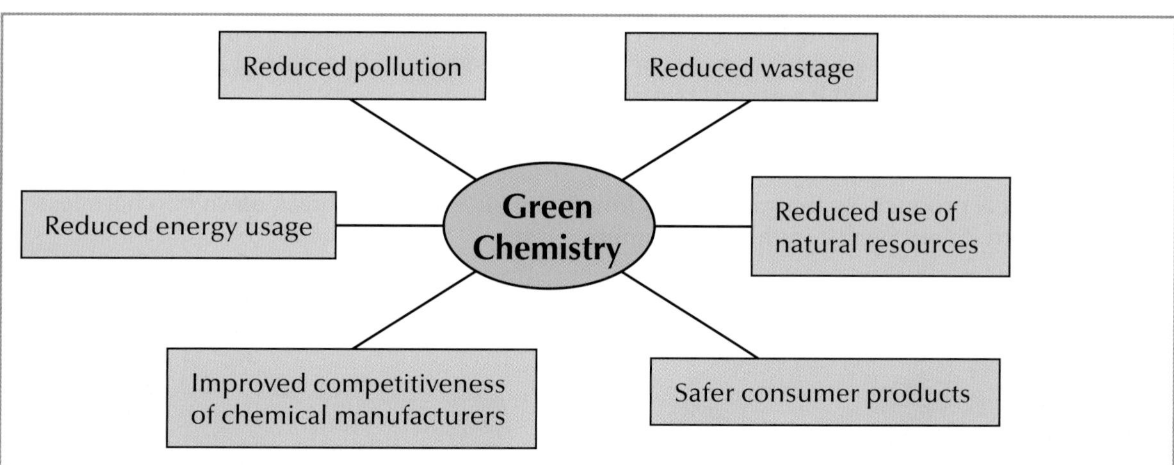

Figure 23.1 *Some benefits of Green Chemistry*

The twelve principles of Green Chemistry

Green Chemistry has **twelve** principles:

1 **Prevent waste**

Design processes that **prevent waste** from being produced rather than having to treat waste or clean it up afterwards.

2 **Maximise atom economy**

Design processes which incorporate most or all of the starting materials into the final products so that **few atoms are wasted**.

3 **Design less hazardous chemical syntheses**

Design processes which use and generate substances which are as **non-toxic** as possible to humans and the environment.

4 **Design safer chemicals and products**

Design chemical products that are as effective as possible whilst being as **non-toxic** as possible.

5 **Use safer solvents and auxiliaries**

Avoid using solvents, separating agents and other additional (auxiliary) chemicals, or replace them with **safer alternatives**.

6 **Increase energy efficiency**

Use the **minimum amount of energy**; whenever possible carry out processes at room temperature and pressure.

7 Use renewable feedstocks

Use raw materials (feedstocks) which are **renewable** and reduce the use of non-renewable raw materials to a minimum.

8 Reduce derivatives

Avoid **derivatisation** because it requires additional reagents and can generate waste, e.g. avoid temporarily modifying physical and chemical processes to prevent unwanted reactions.

9 Use catalysts rather than stoichiometric reagents

Use **catalysts** whenever possible because they are effective in small amounts, can carry out a single reaction many times, cause reactions to occur faster and at lower temperatures, and reduce the production of unwanted and hazardous by-products.

10 Design for degradation

Design chemical products so that when their functional life ends, they **break down** into harmless products which do not persist in the environment.

Biofuels are made from renewable raw materials such as maize and oil palm

Biodegradable plastic bags

11 Analyse in real-time to prevent pollution

Monitor the progress of any process to prevent the formation of any unwanted or hazardous by-products.

12 Minimise the potential for accidents

Choose reagents to be used in chemical processes which keep the possibility of **chemical accidents** to a minimum, e.g. explosions, fires and the release of toxic substances.

Revision questions

1. Define the term 'Green Chemistry'.
2. Give FOUR benefits of Green Chemistry.
3. Identify FOUR principles of Green Chemistry.

24 Qualitative analysis

Qualitative analysis involves **identifying** the components of a single substance or a mixture, for example, identifying the cation and anion present in an ionic compound.

Identifying cations

Reaction with aqueous sodium hydroxide (sodium hydroxide solution)

Metal cations form insoluble metal hydroxides when they react with aqueous sodium hydroxide:

$$M^{n+}(aq) + nOH^{-}(aq) \longrightarrow M(OH)_n(s)$$

To identify the metal or ammonium ion in an unknown ionic compound, make a **solution** of the solid in distilled water. Then add aqueous sodium hydroxide **dropwise** and note the **colour** of any **precipitate** that forms. Add **excess** aqueous sodium hydroxide and look to see if the precipitate **dissolves**. If **no** precipitate is seen, warm gently and test for **ammonia** gas (see Table 24.8, p. 190).

Table 24.1 *Reactions of cations with aqueous sodium hydroxide*

Cation	Colour of precipitate on adding aqueous sodium hydroxide dropwise	Identity of precipitate	Reaction equation	Effect of adding excess aqueous sodium hydroxide (NaOH(aq)) to the precipitate
Ca^{2+}	**White**	**Calcium hydroxide**	$Ca^{2+}(aq) + 2OH^{-}(aq) \longrightarrow Ca(OH)_2(s)$ white	Precipitate remains, i.e. it is **insoluble** in excess NaOH(aq).
Al^{3+}	**White**	**Aluminium hydroxide**	$Al^{3+}(aq) + 3OH^{-}(aq) \longrightarrow Al(OH)_3(s)$ white	Precipitate dissolves forming a colourless solution, i.e. it is **soluble** in excess NaOH(aq).
Pb^{2+}	**White**	**Lead(II) hydroxide**	$Pb^{2+}(aq) + 2OH^{-}(aq) \longrightarrow Pb(OH)_2(s)$ white	Precipitate dissolves forming a colourless solution, i.e. it is **soluble** in excess NaOH(aq).
Zn^{2+}	**White**	**Zinc hydroxide**	$Zn^{2+}(aq) + 2OH^{-}(aq) \longrightarrow Zn(OH)_2(s)$ white	Precipitate dissolves forming a colourless solution, i.e. it is **soluble** in excess NaOH(aq).
Cu^{2+}	**Blue**	**Copper(II) hydroxide**	$Cu^{2+}(aq) + 2OH^{-}(aq) \longrightarrow Cu(OH)_2(s)$ blue	Precipitate remains, i.e. it is **insoluble** in excess NaOH(aq).

Cation	Colour of precipitate on adding aqueous sodium hydroxide dropwise	Identity of precipitate	Reaction equation	Effect of adding excess aqueous sodium hydroxide (NaOH(aq)) to the precipitate
Fe^{2+}	**Green**	**Iron(II) hydroxide**	$Fe^{2+}(aq) + 2OH^-(aq) \longrightarrow Fe(OH)_2(s)$ green	Precipitate remains, i.e. it is **insoluble** in excess NaOH(aq).
Fe^{3+}	**Red-brown**	**Iron(III) hydroxide**	$Fe^{3+}(aq) + 3OH^-(aq) \longrightarrow Fe(OH)_3(s)$ red-brown	Precipitate remains, i.e. it is **insoluble** in excess NaOH(aq).
NH_4^+	**No** precipitate. **Ammonia** is evolved on warming which changes moist red litmus paper blue.	–	$NH_4^+(aq) + OH^-(aq) \longrightarrow NH_3(g) + H_2O(l)$	–

From left to right, precipitates of lead(II) hydroxide, copper(II) hydroxide, iron(II) hydroxide and iron(III) hydroxide

Calcium hydroxide, copper(II) hydroxide, iron(II) hydroxide and **iron(III) hydroxide** are **basic hydroxides** so they do not react with sodium hydroxide. When excess aqueous sodium hydroxide is added to these precipitates they **remain**.

Aluminium hydroxide, lead(II) hydroxide and **zinc hydroxide** are **amphoteric hydroxides**. They react with the excess sodium hydroxide forming soluble salts, hence the precipitates **dissolve**.

Ammonium hydroxide is soluble, therefore, no precipitate forms.

Reaction with aqueous ammonia (ammonium hydroxide solution)

The metal cations, except the Ca^{2+} ion, form insoluble metal hydroxides with ammonium hydroxide solution:

$$M^{n+}(aq) + nOH^-(aq) \longrightarrow M(OH)_n(s)$$

To identify the metal ion in an unknown ionic compound, make a **solution** of the solid in distilled water. Then add aqueous ammonia **dropwise** and note the **colour** of any **precipitate** that forms. Add **excess** aqueous ammonia and look to see if the precipitate **dissolves**.

Table 24.2 *Reactions of cations with aqueous ammonia*

Cation	Colour of precipitate on adding aqueous ammonia dropwise	Identity of precipitate	Reaction equation	Effect of adding excess aqueous ammonia ($NH_4OH(aq)$) to the precipitate
Ca^{2+}	**No** precipitate	–	–	–
Al^{3+}	**White**	**Aluminium hydroxide**	$Al^{3+}(aq) + 3OH^-(aq) \longrightarrow Al(OH)_3(s)$ white	Precipitate remains, i.e. it is **insoluble** in excess $NH_4OH(aq)$.
Pb^{2+}	**White**	**Lead(II) hydroxide**	$Pb^{2+}(aq) + 2OH^-(aq) \longrightarrow Pb(OH)_2(s)$ white	Precipitate remains, i.e. it is **insoluble** in excess $NH_4OH(aq)$.
Zn^{2+}	**White**	**Zinc hydroxide**	$Zn^{2+}(aq) + 2OH^-(aq) \longrightarrow Zn(OH)_2(s)$ white	Precipitate dissolves forming a colourless solution, i.e. it is **soluble** in excess $NH_4OH(aq)$.
Cu^{2+}	**Blue**	**Copper(II) hydroxide**	$Cu^{2+}(aq) + 2OH^-(aq) \longrightarrow Cu(OH)_2(s)$ blue	Precipitate dissolves forming a deep blue solution, i.e. it is **soluble** in excess $NH_4OH(aq)$.
Fe^{2+}	**Green**	**Iron(II) hydroxide**	$Fe^{2+}(aq) + 2OH^-(aq) \longrightarrow Fe(OH)_2(s)$ green	Precipitate remains, i.e. it is **insoluble** in excess $NH_4OH(aq)$.
Fe^{3+}	**Red-brown**	**Iron(III) hydroxide**	$Fe^{3+}(aq) + 3OH^-(aq) \longrightarrow Fe(OH)_3(s)$ red-brown	Precipitate remains, i.e. it is **insoluble** in excess $NH_4OH(aq)$.

Aluminium hydroxide, **lead(II) hydroxide**, **iron(II) hydroxide** and **iron(III) hydroxide** do not react with ammonium hydroxide. When excess aqueous ammonia is added, these precipitates **remain**.

Zinc hydroxide and **copper(II) hydroxide** do react with ammonium hydroxide. They react with the excess aqueous ammonia forming complex soluble salts, hence the precipitates **dissolve**.

To distinguish between the Al^{3+} ion and the Pb^{2+} ion

Putting the results of the previous two tests together, the **Al^{3+} ion** and the **Pb^{2+} ion** are the only cations which cannot be distinguished. To distinguish between these, make a solution of the solid, add a few drops of **potassium iodide solution** and look for the appearance of a **precipitate**:

- **Al^{3+} ion**: **no** precipitate forms.
- **Pb^{2+} ion**: a **bright yellow** precipitate of **lead(II) iodide** forms:

$$Pb^{2+}(aq) + 2I^-(aq) \longrightarrow \underset{\text{yellow}}{PbI_2(s)}$$

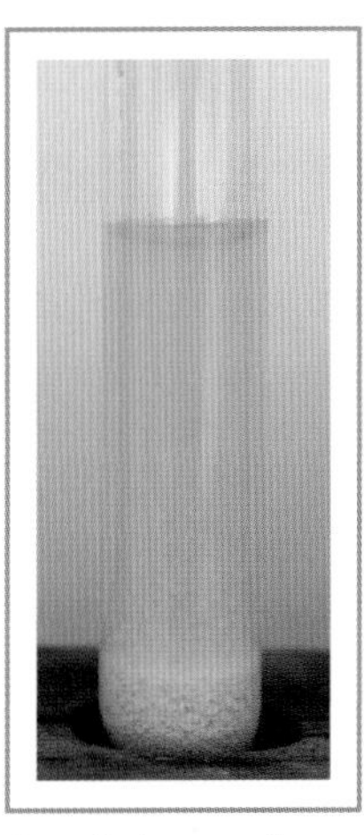

Precipitate of lead(II) iodide

Identifying anions

Effect of heat on the solid

Heat a small quantity of the **solid** in a **dry** test tube and test the **gas** evolved (see Table 24.8, p. 190 for the tests to identify gases).

Table 24.3 *Effect of heat on the solid*

Anion	Observations/results of gas tests	Identity of gas(es) evolved	Reaction equation
CO_3^{2-}	A white precipitate forms in lime water.	**Carbon dioxide**	$CO_3^{2-}(s) \xrightarrow{\text{heat}} O^{2-}(s) + CO_2(g)$
NO_3^- of potassium or sodium	A glowing splint relights.	**Oxygen**	$2NO_3^-(s) \xrightarrow{\text{heat}} 2NO_2^-(s) + O_2(g)$
NO_3^- of calcium and below	A brown gas is seen and a glowing splint relights.	**Nitrogen dioxide** and **oxygen**	$4NO_3^-(s) \xrightarrow{\text{heat}} 2O^{2-}(s) + 4NO_2(g) + O_2(g)$

Further test to identify the nitrate ion

Add concentrated sulfuric acid and a few copper turnings to a sample of the **solid** in a **dry** test tube and heat. A **blue** solution forms and brown **nitrogen dioxide** gas is evolved.

Reaction with dilute acid

Add dilute hydrochloric or nitric acid to a spatula of the **solid** in a test tube, heat if necessary, and test the **gas** evolved.

Table 24.4 *Reactions of anions with dilute acid*

Anion	Results of gas tests	Identity of gas evolved	Reaction equation
CO_3^{2-}	A white precipitate forms in lime water.	**Carbon dioxide**	$CO_3^{2-}(s) + 2H^+(aq) \longrightarrow CO_2(g) + H_2O(l)$
SO_3^{2-}	Acidified potassium manganate(VII) solution changes from purple to colourless.	**Sulfur dioxide**	$SO_3^{2-}(s) + 2H^+(aq) \longrightarrow SO_2(g) + H_2O(l)$

Reaction with concentrated sulfuric acid

Add a few drops of concentrated sulfuric acid to a small amount of the **solid** in a dry test tube and test the **gas** evolved.

Table 24.5 *Reactions of anions with concentrated sulfuric acid*

Anion	Observations/results of gas tests	Identity of gas evolved/product
CO_3^{2-}	A white precipitate forms in lime water.	**Carbon dioxide**
SO_3^{2-}	Acidified potassium manganate(VII) solution changes from purple to colourless.	**Sulfur dioxide**
Cl^-	White fumes form with ammonia gas.	**Hydrogen chloride**
Br^-	A red-brown gas is seen.	**Bromine**
I^-	A grey-black solid forms which sublimes to form a purple vapour if heated.	**Iodine**

Reaction with silver nitrate solution followed by aqueous ammonia

Make a **solution** of the solid in dilute nitric acid. Add a few drops of silver nitrate solution and observe the **colour** of the precipitate. Add aqueous ammonia and look to see if the precipitate **dissolves**.

Table 24.6 *Reactions of anions with silver nitrate solution followed by aqueous ammonia*

Anion	Observations when silver nitrate solution is added	Identity of precipitate	Reaction equation	Observation when aqueous ammonia is added
Cl^-	A **white** precipitate forms which becomes slightly purple-grey in sunlight.	**Silver chloride**	$Ag^+(aq) + Cl^-(aq) \longrightarrow AgCl(s)$ white	Precipitate dissolves forming a colourless solution, i.e. it is **soluble** in aqueous ammonia.
Br^-	A **cream** precipitate forms which becomes slightly green in sunlight.	**Silver bromide**	$Ag^+(aq) + Br^-(aq) \longrightarrow AgBr(s)$ cream	Precipitate partially dissolves, i.e. it is **slightly soluble.**
I^-	A **pale yellow** precipitate forms.	**Silver iodide**	$Ag^+(aq) + I^-(aq) \longrightarrow AgI(s)$ pale yellow	Precipitate remains, i.e. it is **insoluble.**

Reaction with barium nitrate solution followed by dilute acid

To identify the **sulfate ion**, make a **solution** of the solid in distilled water. Add a few drops of barium nitrate (or barium chloride) solution and observe the precipitate. Add dilute nitric or hydrochloric acid, heating if necessary, and look to see if the precipitate **dissolves.** Test any **gas** evolved.

Table 24.7 *Reactions of anions with barium nitrate solution followed by dilute acid*

Anion	Observations when barium nitrate solution is added	Identity of precipitate	Reaction equation	Observations when dilute acid is added
CO_3^{2-}	A **white** precipitate forms.	**Barium carbonate**	$Ba^{2+}(aq) + CO_3^{2-}(aq) \longrightarrow BaCO_3(s)$ white	Precipitate **dissolves;** it reacts releasing **carbon dioxide:** $CO_3^{2-}(s) + 2H^+(aq) \longrightarrow CO_2(g) + H_2O(l)$
SO_3^{2-}	A **white** precipitate forms.	**Barium sulfite**	$Ba^{2+}(aq) + SO_3^{2-}(aq) \longrightarrow BaSO_3(s)$ white	Precipitate **dissolves;** it reacts releasing **sulfur dioxide** on heating: $SO_3^{2-}(s) + 2H^+(aq) \longrightarrow SO_2(g) + H_2O(l)$
SO_4^{2-}	A **white** precipitate forms.	**Barium sulfate**	$Ba^{2+}(aq) + SO_4^{2-}(aq) \longrightarrow BaSO_4(s)$ white	Precipitate **remains;** it does not react with dilute acid.

Identifying gases

A variety of tests can be performed in the laboratory to identify different **gases**.

Table 24.8 *Identifying gases*

Gas	Properties	Test	Explanation of test
Oxygen (O_2)	Colourless and odourless.	• Causes a **glowing splint** to glow brighter or **relight**.	• Oxygen supports combustion.
Hydrogen (H_2)	Colourless and odourless.	• Causes a **lighted splint** to make a '**squeaky pop**' and be extinguished.	• Hydrogen reacts explosively with oxygen in the air to form steam: $2H_2(g) + O_2(g) \longrightarrow 2H_2O(g)$
Carbon dioxide (CO_2)	Colourless and odourless.	• Forms a **white precipitate** in **lime water** (calcium hydroxide solution). • The precipitate **redissolves** on continued bubbling.	• Insoluble, white calcium carbonate forms: $Ca(OH)_2(aq) + CO_2(g) \longrightarrow \underset{\text{white}}{CaCO_3(s)} + H_2O(l)$ • Soluble calcium hydrogencarbonate forms: $CaCO_3(s) + H_2O(l) + CO_2(g) \longrightarrow Ca(HCO_3)_2(aq)$
Hydrogen chloride (HCl)	Colourless, with sharp, acid odour.	• Forms **white fumes** when brought near a drop of **concentrated ammonia solution** on a glass rod.	• White fumes of ammonium chloride form: $NH_3(g) + HCl(g) \longrightarrow NH_4Cl(s)$
Ammonia (NH_3)	Colourless, with pungent odour.	• Turns **moist red litmus paper blue**. • Forms **white fumes** when brought near a drop of **concentrated hydrochloric acid** on a glass rod.	• Ammonia reacts with water on the paper forming alkaline ammonium hydroxide: $NH_3(g) + H_2O(l) \rightleftharpoons NH_4OH(aq)$ • White fumes of ammonium chloride form: $NH_3(g) + HCl(g) \longrightarrow NH_4Cl(s)$
Chlorine (Cl_2)	Yellow-green, with sharp, choking odour.	• Causes **moist blue litmus paper** to turn **red** and then bleaches it **white**.	• Chlorine reacts with water on the paper forming two acids, hydrochloric acid (HCl) and chloric(I) acid (HClO): $Cl_2(g) + H_2O(l) \longrightarrow HCl(aq) + HClO(aq)$ Chloric(I) acid oxidises coloured litmus to colourless.

Gas	Properties	Test	Explanation of test
Sulfur dioxide (SO_2)	Colourless, with pungent odour.	• Turns **acidified potassium manganate(VII) solution** from **purple** to **colourless**. • Turns **acidified potassium dichromate(VI) solution** from **orange** to **green**.	• Sulfur dioxide reduces the purple MnO_4^- ion to the colourless Mn^{2+} ion. • Sulfur dioxide reduces the orange $Cr_2O_7^{2-}$ ion to the green Cr^{3+} ion.
Nitrogen dioxide (NO_2)	Brown, with sharp, irritating odour.	• A **brown gas**. • Causes **moist blue litmus paper** to turn **red**.	• Nitrogen dioxide reacts with water on the paper forming two acids, nitrous acid (HNO_2) and nitric acid (HNO_3): $2NO_2(g) + H_2O(l) \longrightarrow HNO_2(aq) + HNO_3(aq)$
Water vapour (H_2O)	Colourless and odourless.	• Causes **dry cobalt(II) chloride paper** to change from **blue** to **pink**. • Causes **anhydrous copper(II) sulfate** to change from **white** to **blue**.	• Water vapour changes blue, anhydrous cobalt(II) chloride ($CoCl_2$) to pink, hydrated cobalt(II) chloride ($CoCl_2.6H_2O$). • Water vapour changes white, anhydrous copper(II) sulfate ($CuSO_4$) to blue, hydrated copper(II) sulfate ($CuSO_4.5H_2O$).

Revision questions

1. A few drops of aqueous sodium hydroxide were added to a solution containing aluminium ions. A white precipitate formed and then dissolved when excess aqueous sodium hydroxide was added.

   a Explain why the precipitate formed and write an ionic equation for its formation.

   b Why did the precipitate dissolve in excess aqueous ammonia?

2. A green precipitate formed when a few drops of aqueous ammonia were added to a solution of an unknown ionic compound **X**, and the precipitate remained when excess aqueous ammonia was added.

   a Suggest the identity of the cation in **X**.

   b Explain why the precipitate formed and write an ionic equation for its formation.

   c Why did the precipitate remain in excess aqueous ammonia?

3. A solid sample of an ionic compound **Y** was heated in a test tube and a brown gas was evolved and a glowing splint relit when placed in the tube.

   a Suggest the identity of the gases evolved.

   b Suggest the identity of the anion in **Y**.

4 Dilute hydrochloric acid was added to a sample of ionic compound **Z** and the gas evolved caused acidified potassium manganate(VII) solution to change from purple to colourless.

a Suggest the identity of the gas evolved.

b Suggest the identity of the anion in **Z**.

c Explain the colour change in the acidified potassium manganate(VII) solution.

5 For EACH of the following pairs of ionic compounds, describe ONE test you could use to distinguish between them. Your answer should include the results of the test on BOTH substances in EACH pair and relevant ionic equations.

a lead(II) nitrate and zinc nitrate

b lead(II) nitrate and aluminium nitrate

c potassium chloride and potassium iodide

d sodium sulfate and sodium sulfite

6 Describe how you would test for the presence of EACH of the following gases

a hydrogen

b ammonia

c chlorine

d water vapour

7 An unknown gas **W** was bubbled into limewater and a white precipitate formed. Further bubbling of the gas caused the limewater to become colourless once more. Suggest an identity for **W** and account for the observations. Support your answer with the relevant equations.

Exam-style questions – Chapters 18 to 24

Structured questions

1 a) Depending on conditions, a solid non-metallic element, X, burns in oxygen to form one of two oxides, Y or Z. Both Y and Z are colourless gases.

– Y is capable of reacting with water to form a solution that turns blue litmus red.

– Z does not react with water, dilute acids or alkalis.

i) Identify X, Y and Z. **(3 marks)**

ii) What factor determines whether X forms Y or Z? **(1 mark)**

b) i) Write a balanced equation for the reaction between magnesium and hydrogen. **(1 mark)**

ii) Is hydrogen behaving as an oxidising agent or a reducing agent in the reaction in **i)** above? Give a reason for your answer. **(2 marks)**

c) Joy carried out some tests on a solid labelled R. Some of her observations and inferences are recorded in Table 1 below. Complete the table by filling in the other observations and inferences.

Table 1 *Observations and inferences from tests on R*

Test	Observations	Inferences
A sample of R was heated strongly in a dry test tube and a glowing splint was placed in the mouth of the tube.	• Brown fumes were seen. • The glowing splint relit.	• • • **(3 marks)**
Aqueous sodium hydroxide was added to a solution of R until in excess.	• A white precipitate formed which dissolved in excess to form a colourless solution.	• **(1 mark)**
Aqueous ammonia was added to a solution of R until in excess.	• A white precipitate formed which dissolved in excess to form a colourless solution.	• Ionic equation required: **(2 marks)**
A solution of R was made in dilute nitric acid and a few drops of silver nitrate solution were added followed by aqueous ammonia.	• **(1 mark)**	• Cl^- ions present. Ionic equation required: **(1 mark)**

Total 15 marks

2 **a)** Figure 1 below shows the apparatus that Tom assembled to prepare and collect ammonia in the laboratory.

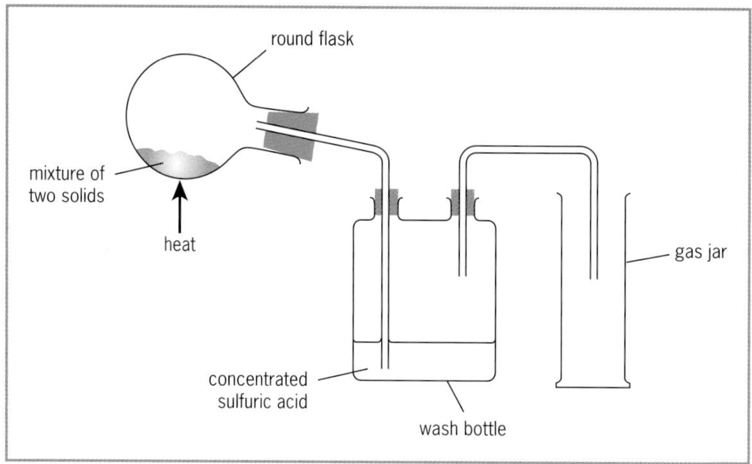

Figure 1 *Apparatus used to make ammonia in the laboratory*

i) Name TWO suitable solids that Tom could have put into the round flask. **(1 mark)**

ii) Based on the information in Figure 1, suggest TWO reasons why Tom's method was unsuccessful and he was unable to collect a jar of ammonia. **(2 marks)**

iii) Identify ONE test Tom could use to prove that the gas being produced was ammonia. Include a chemical equation to represent the test you have described. **(3 marks)**

b) Carbon dioxide, which has many uses in today's world as well as posing a threat to the environment, can also be prepared in the laboratory.

i) Identify TWO reactants that would be suitable for the preparation. **(1 mark)**

ii) Give ONE use of carbon dioxide and ONE property that makes it suitable for the use identified. **(2 marks)**

iii) Identify ONE activity of man which is causing an increase in the concentration of carbon dioxide in the environment. **(1 mark)**

iv) Suggest TWO consequences of the build-up of carbon dioxide which is occurring in the upper atmosphere. **(2 marks)**

c) Explain why the overuse of chemical fertilisers on agricultural land poses a threat to aquatic environments, particularly lakes and ponds. **(3 marks)**

Total 15 marks

Extended response questions

3 **a)** Experiments carried out on TWO metals, P and Q, gave the following results:

- P displaced silver from a solution of silver nitrate but did not react with iron(III) nitrate solution.
- Q reacted vigorously with cold water.

i) Based on these results and your knowledge of the reactivity series of metals, place metals P, Q, silver, aluminium, magnesium and iron in decreasing order of reactivity. **(3 marks)**

ii) If P forms a divalent cation and Q forms a monovalent cation, write equations for:

– the reaction between P and silver nitrate solution.

– the reaction between Q and water.

Use P and Q as symbols of the elements, you are not expected to identify them. **(4 marks)**

iii) Suggest, with a reason, the most suitable method for extracting Q from its ore. **(2 marks)**

b) i) Outline the main steps in the extraction of iron from its ores. **(4 marks)**

ii) Stainless steel, an alloy of iron, is used extensively to make cooking utensils. Suggest ONE reason why it is used instead of iron for this purpose. **(1 mark)**

iii) State ONE other use of iron or its alloys. **(1 mark)**

Total 15 marks

4 **a)** Using the principles for placing metals in the reactivity series, explain why copper, silver and gold are used for making coins, but magnesium, sodium and calcium are not used. **(4 marks)**

b) How would the effect of heat on sodium nitrate differ from the effect of heat on lead(II) nitrate? Support your answer with relevant equations. **(4 marks)**

c) Certain metal ions play vital roles in the life of living organisms whilst others can be extremely harmful and cause serious pollution.

i) What is meant by the term 'pollution'? **(1 mark)**

ii) Outline the roles played by

– magnesium

– iron

in the lives of plants and animals. **(4 marks)**

iii) Explain why it can be potentially dangerous to consume large quantities of fish such as marlin, tuna and shark. **(2 marks)**

Total 15 marks

5 **a)** Scientists believe that life would cease to exist without water. Outline THREE properties of water and relate them to its function in sustaining life on Earth. **(6 marks)**

b) You went to visit your friend who lives in another country and noticed that the soap you usually use did not lather well and produced scum instead.

i) Explain the reason why your soap produced scum when you tried to make it lather. **(3 marks)**

ii) Describe ONE way that your friend could treat her water to prevent it from forming scum. Include a chemical equation in your answer. **(2 marks)**

c) i) What is meant by 'Green Chemistry'? **(1 mark)**

ii) Outline TWO of the principles of Green Chemistry. **(2 marks)**

iii) Suggest ONE benefit gained by implementing the principles of Green Chemistry. **(1 mark)**

Total 15 marks

Appendix – The Periodic Table

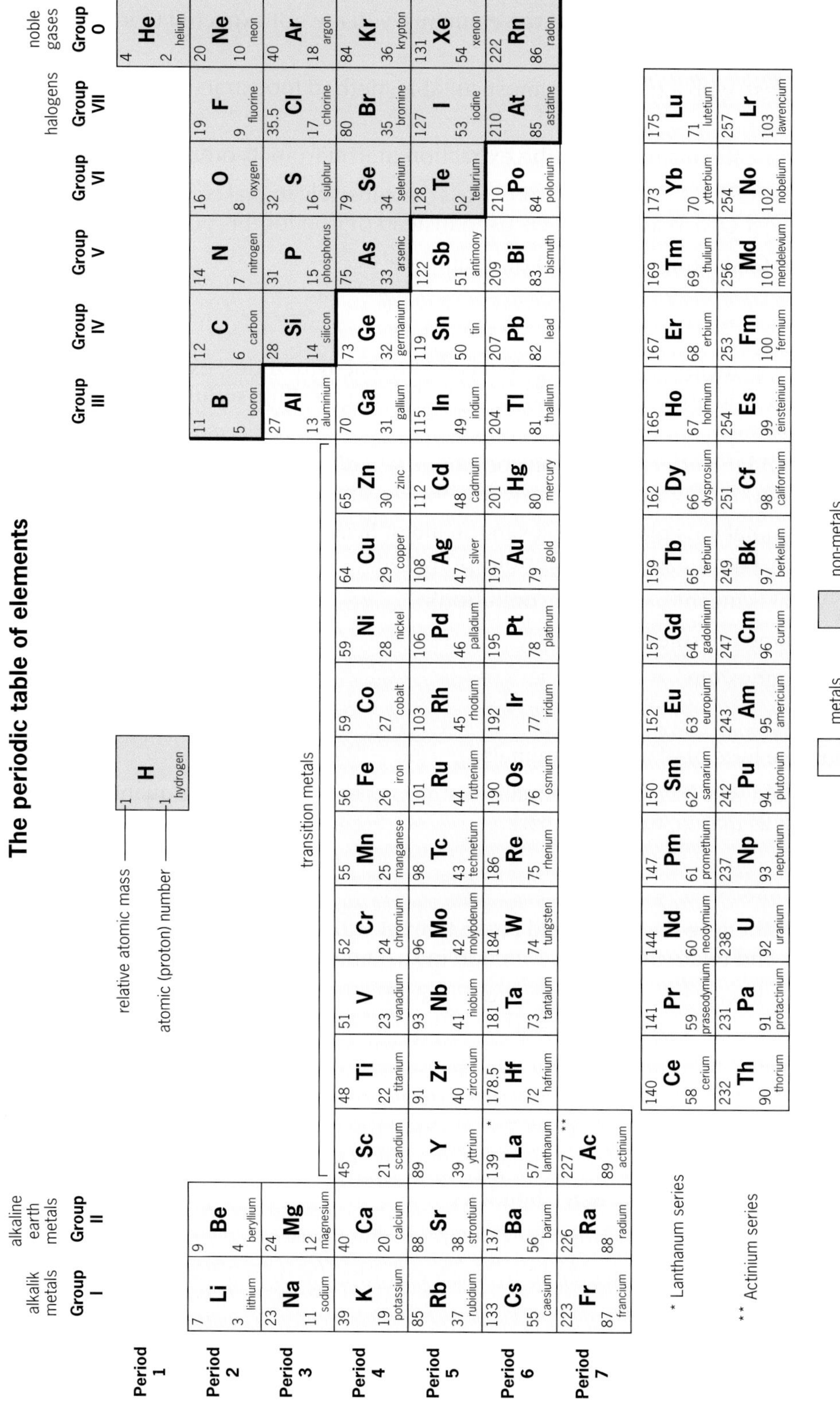

The periodic table of elements

relative atomic mass — 1; atomic (proton) number — 1; **H** hydrogen

	Group I (alkalik metals)	Group II (alkaline earth metals)	transition metals										Group III	Group IV	Group V	Group VI	Group VII (halogens)	Group 0 (noble gases)
Period 1																		4 **He** 2 helium
Period 2	7 **Li** 3 lithium	9 **Be** 4 beryllium											11 **B** 5 boron	12 **C** 6 carbon	14 **N** 7 nitrogen	16 **O** 8 oxygen	19 **F** 9 fluorine	20 **Ne** 10 neon
Period 3	23 **Na** 11 sodium	24 **Mg** 12 magnesium											27 **Al** 13 aluminium	28 **Si** 14 silicon	31 **P** 15 phosphorus	32 **S** 16 sulphur	35.5 **Cl** 17 chlorine	40 **Ar** 18 argon
Period 4	39 **K** 19 potassium	40 **Ca** 20 calcium	45 **Sc** 21 scandium	48 **Ti** 22 titanium	51 **V** 23 vanadium	52 **Cr** 24 chromium	55 **Mn** 25 manganese	56 **Fe** 26 iron	59 **Co** 27 cobalt	59 **Ni** 28 nickel	64 **Cu** 29 copper	65 **Zn** 30 zinc	70 **Ga** 31 gallium	73 **Ge** 32 germanium	75 **As** 33 arsenic	79 **Se** 34 selenium	80 **Br** 35 bromine	84 **Kr** 36 krypton
Period 5	85 **Rb** 37 rubidium	88 **Sr** 38 strontium	89 **Y** 39 yttrium	91 **Zr** 40 zirconium	93 **Nb** 41 niobium	96 **Mo** 42 molybdenum	98 **Tc** 43 technetium	101 **Ru** 44 ruthenium	103 **Rh** 45 rhodium	106 **Pd** 46 palladium	108 **Ag** 47 silver	112 **Cd** 48 cadmium	115 **In** 49 indium	119 **Sn** 50 tin	122 **Sb** 51 antimony	128 **Te** 52 tellurium	127 **I** 53 iodine	131 **Xe** 54 xenon
Period 6	133 **Cs** 55 caesium	137 **Ba** 56 barium	139 * **La** 57 lanthanum	178.5 **Hf** 72 hafnium	181 **Ta** 73 tantalum	184 **W** 74 tungsten	186 **Re** 75 rhenium	190 **Os** 76 osmium	192 **Ir** 77 iridium	195 **Pt** 78 platinum	197 **Au** 79 gold	201 **Hg** 80 mercury	204 **Tl** 81 thallium	207 **Pb** 82 lead	209 **Bi** 83 bismuth	210 **Po** 84 polonium	210 **At** 85 astatine	222 **Rn** 86 radon
Period 7	223 **Fr** 87 francium	226 **Ra** 88 radium	227 ** **Ac** 89 actinium															

* Lanthanum series	140 **Ce** 58 cerium	141 **Pr** 59 praseodymium	144 **Nd** 60 neodymium	147 **Pm** 61 promethium	150 **Sm** 62 samarium	152 **Eu** 63 europium	157 **Gd** 64 gadolinium	159 **Tb** 65 terbium	162 **Dy** 66 dysprosium	165 **Ho** 67 holmium	167 **Er** 68 erbium	169 **Tm** 69 thulium	173 **Yb** 70 ytterbium	175 **Lu** 71 lutetium
** Actinium series	232 **Th** 90 thorium	231 **Pa** 91 protactinium	238 **U** 92 uranium	237 **Np** 93 neptunium	242 **Pu** 94 plutonium	243 **Am** 95 americium	247 **Cm** 96 curium	249 **Bk** 97 berkelium	251 **Cf** 98 californium	254 **Es** 99 einsteinium	253 **Fm** 100 fermium	256 **Md** 101 mendelevium	254 **No** 102 nobelium	257 **Lr** 103 lawrencium

metals ▢ non-metals (shaded)

Index